Therapeutics and Human Physiology

PHARMACY PRACTICE

edited by JASON HALL

INTEGRATED FOUNDATIONS of PHARMACY

THERAPEUTICS AND
HUMAN PHYSIOLOGY

how drugs work

edited by ELSIE GASKELL & CHRIS ROSTRON

INTEGRATED FOUNDATIONS of PHARMACY

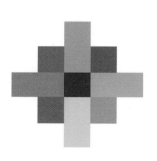

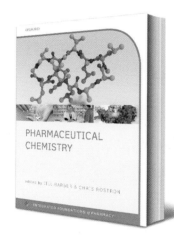

PHARMACEUTICAL
CHEMISTRY

edited by JILL BARBER & CHRIS ROSTRON

INTEGRATED FOUNDATIONS of PHARMACY

PHARMACEUTICS

the science of medicine design

edited by PHILIP DENTON & CHRIS ROSTRON

INTEGRATED FOUNDATIONS of PHARMACY

INTEGRATED FOUNDATIONS *of* PHARMACY

Therapeutics and Human Physiology

How Drugs Work

Edited by Elsie E. Gaskell and Chris Rostron

OXFORD

UNIVERSITY PRESS

 INTEGRATED FOUNDATIONS *of* PHARMACY

OXFORD
UNIVERSITY PRESS

Great Clarendon Street, Oxford, OX2 6DP,
United Kingdom

Oxford University Press is a department of the University of Oxford.
It furthers the University's objective of excellence in research, scholarship,
and education by publishing worldwide. Oxford is a registered trade mark of
Oxford University Press in the UK and in certain other countries

British Library Cataloguing in Publication Data

Data available

ISBN 978–0–19–965529–8

Printed in Italy by
L.E.G.O. S.p.A.—Lavis TN

Preface

As a result of significant changes taking place within the profession of pharmacy, there is an increasing trend for universities to adopt an integrated approach to pharmacy education. There is now an overwhelming view that pharmaceutical science must be combined with the more practice-oriented aspects of pharmacy. This assists students to see the impact of, and relationships between, those subjects which make up the essential knowledge base for a practising pharmacist.

This series supports integrated pharmacy education, so that from day one of the course students can take a professionally relevant approach to their learning. This is achieved through the organization of content and the use of key learning features. Cross-references highlight related topics of importance, directing the reader to further information; Integration Boxes flag relevant topics in a different strand of pharmacy; and Case Studies explore the ways in which pharmaceutical science and practice impacts upon patients' lives, allowing material to be addressed in a patient-centred context.

There are four books in the series, covering each main strand: *Pharmaceutical Chemistry, Pharmaceutics: The Science of Medicine Design, Pharmacy Practice*, and *Therapeutics and Human Physiology: How Drugs Work*. Each book is edited by a subject expert, with contributors from across pharmacy education. They have been carefully written to ensure an appropriate breadth and depth of knowledge for the first year student.

Each book concludes with an overview of the subject and application to pharmacy, building on students' understanding of the concepts and bringing it all together. It applies the material to pharmacy practice in a variety of ways and places it in the context of all four pharmacy strands.

Therapeutics and Human Physiology: How Drugs Work

Our bodies are complex and integrated systems and only with an integrated approach to understanding how they work and linking this to related pharmaceutical sciences and practice-oriented disciplines will the way drugs work be fully appreciated. *Therapeutics and Human Physiology: How Drugs* Work spans a variety of areas and introduces fundamental concepts of therapeutics together with relevant aspects of cell and molecular biology, biochemistry, human physiology and anatomy, and pharmacology. Although the book is written as distinct chapters, it should provide a sound foundation upon which future learning can be built on and equip the reader with skills to identify and bring together related topics across the different disciplines not only associated with therapeutics but pharmacy as a whole.

Knowledge in the area of therapeutics and current best practice are continuously changing. The text within this book contains drug names and relevant doses as currently prescribed. Readers should always consult the most up-to-date source for guidance on drug therapies and doses used.

Elsie Gaskell and Chris Rostron, July 2012

Acknowledgements

The editors wish to acknowledge the support of all the contributing authors who have spent a considerable amount of time writing and reviewing their chapters. We would also like to extend our thanks to the staff at Oxford University Press, especially Holly Edmundson (Publishing Editor) for her continued support, encouragement and guidance. In addition we are grateful to Jonathan Crowe, Jackie Parker, Philippa Hendry, and Erica Martin for their assistance with preparing the manuscripts.

The supportive feedback from the reviewers: Dr Stephen Kelley, Dr Gurprit Lall, Dr David Poyner, Dr Julie Sanderson, and Dr Helen Vosper, was greatly appreciated.

Finally we would like to thank our past, present, and future students who have inspired the production of this book.

Contents

An introduction to the *Integrated Foundations of Pharmacy* series

The path to becoming a qualified pharmacist is incredibly rewarding, but it requires diligence. Not only will you need to assimilate knowledge from a range of disciplines, but you must also understand—and demonstrate—how to apply this knowledge in a practical, hands-on environment. This series is written to support you as you encounter the challenges of pharmacy education today.

There are a range of features used in the series, each carefully designed to help you master the material and to encourage to you to see the connections between the different strands of the discipline.

Mastering the material

Boxes

Additional material that adds interest or depth to concepts covered in the main text is provided in the Boxes.

> **BOX 2.1**
>
> **Bacteria versus archaea**
>
> Archaea are often found living in some of the most extreme environments found on Earth, in conditions where humans would not be able to survive. These

Key points

The important 'take home messages' that you must have a good grasp of are highlighted in the Key points. You may find these form a helpful basis for your revision.

> **KEY POINT**
>
> All our cells contain identical genetic information; however the cells differentiate to make up the various types of tissues, organs, and body systems that we are made of.

Self check questions

Questions are provided throughout the chapters in order for you to test your understanding of the material. Take the time to complete these, as they will allow you to evaluate how you are getting on, and they will undoubtedly aid your learning. Answers are provided at the back of each volume.

> **SELF CHECK 1.2**
>
> Why may it be important for health care professionals to establish the family medical history?

Further reading

In this section, we direct you to additional resources that we encourage you to seek out, in your library or online. They will help you to gain a deeper understanding of the material presented in the text.

> **FURTHER READING**
>
> Boarder, M., Newby, D., and Navti, P. *Pharmacology for Pharmacy and the Health Sciences: A Patient-Centred Approach.* Oxford University Press, 2010.

Glossary

You will need to master a huge amount of new terminology as you study pharmacy. The glossaries in each volume should help you with this. Glossary terms are shown in pink.

> **ACE inhibitor** Drug that inhibits angiotensin-converting enzyme.
>
> **Acetylcholine (Ach)** The neurotransmitter at preganglionic autonomic neurons and at postganglionic parasympathetic neurons and at the neuromuscular junction. It acts on nicotinic and muscarinic receptors. Unusually, acetylcholine is also released from sympathetic nerves that

Online resources

Visit the Online Resource Centre for related materials, including 10 multiple choice questions for each chapter, with answers and feedback.

Go to: **www.oxfordtextbooks.co.uk/orc/ifp**

Seeing the connections

Integration boxes

Examples in these Integration boxes show how pharmacy practice, pharmaceutics, pharmaceutical chemistry, human physiology, and therapeutics are all closely interlinked.

INTEGRATION BOX 8.2

Fever

Body temperature is often raised in response to a bacterial or viral infection. This increased core temperature

Case studies

Case studies show how the science you learn at university will impact on how you might advise a patient. Reflection questions and sample answers encourage you to think critically about the points raised in the case study.

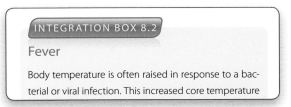
CASE STUDY 2.1

Angela has told Ravi she is pregnant and they are both thrilled at the prospect of being parents. However, Angela had an older brother who had cystic fibrosis (CF)

Cross-references

Linking related sections across all four volumes in the series (as well as other sections within this volume), cross-references will give you a good idea of just how integrated the subject is. Importantly, it will allow you to easily access material on the same subject, as viewed from the perspectives of the different strands of the discipline.

 The study of dosage forms is covered in the **Pharmaceutics** book within this series.

Lecturer support materials

For registered adopters of the volumes in this series, the Online Resource Centre also features figures in electronic format, available to download, for use in lecture presentations and other educational resources.

To register as an adopter, visit www.oxfordtextbooks.co.uk/orc/ifp, select the volume you are interested in, and follow the on-screen instructions.

Any comments?

We welcome comments and feedback about any aspect of the series. Just visit www.oxfordtextbooks.co.uk/orc/feedback and share your views.

About the editors

Editor, Dr Elsie E. Gaskell, studied applied microbiology at Liverpool John Moores University. She worked at Pfizer's Veterinary Medicine R & D department, in Sandwich, Kent, before undertaking a research degree. She completed her PhD in 2006, working in the area of novel therapeutic systems for cystic fibrosis. She is now a senior lecturer in the School of Pharmacy and Biomolecular Sciences at Liverpool John Moores University, where she teaches in the area of pharmacology. Her research interests remain in drug action and delivery.

Series Editor, Dr Chris Rostron, graduated in pharmacy from Manchester University and completed a PhD in medicinal chemistry at Aston University. He gained Chartered Chemist status in 1975. After a period of postdoctoral research he was appointed lecturer in medicinal chemistry at Liverpool Polytechnic. He is now an Honorary Research Fellow in the School of Pharmacy and Biomolecular Sciences at Liverpool John Moores University. Prior to this he was a Reader in Medicinal Chemistry and the Academic Manager at the school. He was a member of the Academic Pharmacy Group Committee of the Royal Pharmaceutical Society of Great Britain and chairman for the past 5 years. He is currently chairman of the Academic Pharmacy Forum and deputy chair of the Education Expert Advisory Panel of the Royal Pharmaceutical Society. He is a past and present external examiner in medicinal chemistry at a number of schools of pharmacy both in the UK and abroad. In 2008 he was awarded honorary membership of the Royal Pharmaceutical Society of Great Britain for services to pharmacy education.

Contributors

Dr Rachel Airley, Division of Pharmacy and Pharmaceutical Sciences, University of Huddersfield, UK

Dr Gillian L. Allison, School of Pharmacy, University of the West Indies, Trinidad and Tobago

Dr Helen Burrell, School of Pharmacy and Biomolecular Sciences, Liverpool John Moores University, UK

Dr Peter N. C. Elliott, School of Pharmacy and Biomolecular Sciences, Liverpool John Moores University, UK

Dr Andrew Evans, School of Pharmacy and Biomolecular Sciences, Liverpool John Moores University, UK

Dr William Ford, Cardiff School of Pharmacy and Pharmaceutical Sciences, Cardiff University, UK

Dr Neil Henney, School of Pharmacy and Biomolecular Sciences, Liverpool John Moores University, UK

Dr Emma Lane, Cardiff School of Pharmacy and Pharmaceutical Sciences, Cardiff University, UK

Dr Peter Penson, School of Pharmacy and Biomolecular Sciences, Liverpool John Moores University, UK

Abbreviations

1,25-DHC	1,25-dihydroxycholecalciferol	EC_{50}	Effective concentration at 50% of E_{max}
5FU	Fluorouracil	E_{max}	Maximal response
5-HIAA	5-hydroxyindoleacetic acid	EPP	Endplate potential
5-HT	5-hydroxytryptamine	EPSP	Excitatory postsynaptic potential
A	Adenine	ERT	Enzyme replacement therapy
ACE	Angiotensin-converting enzyme	ES	Enzyme–substrate complex
ACh	Acetylcholine	ESI	Enzyme–substrate–inhibitor complex
ACTH	Adrenocorticotropic hormone	FAD	Flavin adenine dinucleotide
ADA	Adenosine deaminase	FSH	Follicle-stimulating hormone
ADH	Antidiuretic hormone	G	Guanine
ADHD	Attention deficit hyperactivity disorder	G1	Gap 1 phase of the cell cycle
ADME	Absorption, distribution, metabolism and excretion	G2	Gap 2 phase of the cell cycle
		GABA	γ-aminobutyric acid
ADP	Adenosine-5′-diphosphate	GDP	Guanosine-5′-diphosphate
ADR	Adverse drug reaction	GH	Growth hormone
ALL	Acute lymphoblastic leukaemia	GI	Gastrointestinal
AML	Acute myeloid leukaemia	GLP-1	Glucagon-like peptide-1 agonists
AMP	Adenosine-5′-monophosphate	GTP	Guanosine-5′-triphosphate
ANS	Autonomic nervous system	Hb	Haemoglobin
ATP	Adenosine-5′-triphosphate	HbS	Haemoglobin S
BMR	Basal metabolic rate	HCT	Haematocrit
BNF	British National Formulary	HPV	Human papilloma virus
BP	Blood pressure	HSC	Haematopoietic stem cell
C	Cytosine	IBS	Irritable bowel syndrome
cAMP	3′,5′-Cyclic adenosine monophosphate	ICSH	Intestitial cell-stimulating hormone
CF	Cystic fibrosis	IgE	Immunoglobulin E
CFTR	Cystic fibrosis transmembrane regulator	IL-1	Interleukin 1
		IL-6	Interleukin 6
cGMP	3′,5′-Cyclic guanosine monophosphate	Inr	Initiator
Cl	Clearance	IP_3	Inositol 1,4,5-trisphosphate
CLL	Chronic lymphocytic leukaemia	IPSP	Inhibitory postsynaptic potential
CML	Chronic myeloid leukaemia	k	Rate constant for an enzymatic reaction
CNS	Central nervous system		
CoA	Coenzyme A	k_m	Michaelis constant
COX	Cyclo-oxygenase (COX1 and COX2)	LA	Local anaesthetic
CRH	Corticotropin-releasing hormone	L-DOPA	L-dihydroxyphenylalanine
DAG	Diacylglycerol	LH	Luteinizing hormone
DNA	Deoxyribonucleic acid	LT	Leukotrienes (LTB_4, LTC_4, LTD_4, LTE_4)
DPP-4	Dipeptidyl peptidase-4	M	Mitosis phase of the cell cycle
E	Enzyme	mAChR	Muscarinic acetylcholine receptor
EC	Enzyme Commission	Met	Methionine

miRNA	MicroRNA
MCH	Mean corpuscular haemoglobin
MCV	Mean corpuscular volume
mRNA	Messenger RNA
nAChR	Nicotinic acetylcholine receptor
NAD$^+$	Nicotinamide adenine dinucleotide
NADP$^+$	Nicotinamide adenine dinucleotide phosphate
NANC	Non-adrenergic non-cholinergic control
NK	Natural killer
NSAID	Non-steroidal anti-inflammatory drug
ORC	Origin recognition complex
P	Product
P53	Tumour protein 53
PG	Prostaglandins (PGE$_1$, PGE$_2$, PGI$_2$, PGD$_2$, PGF$_{2\alpha}$)
PIP$_2$	Phosphatidylinositol 4,5-bisphosphate
PMN	Polymorphonuclear
PNS	Peripheral nervous system
Pol	A family of eukaryotic DNA polymerises (α, δ, ε)
PVN	Paraventricular nucleus
RA	Rheumatoid arthritis
RBC	Red blood cell
Rh	Rhesus
RNA	Ribonucleic acid
RNA Pol II	RNA polymerase II
RNAi	RNA interference
rRNA	Ribosomal RNA
[S]	Substrate concentration

S	Synthesis phase of the cell cycle
SA	Sinoatrial
SCA	Sickle cell anaemia
SCID	Severe combined immunodeficiency disease
SI	International System of Units (Système International d'Unités)
siRNA	Short interfering
snRNA	Small nuclear RNA
SSRI	Selective serotonin reuptake inhibitor
STAT	Signal transducer and activator of transcription
T	Thymine
$t_{1/2}$	Half-life
THC	Tetrahydrocannabinol
TK	Tyrosine kinase
TNF-α	Tumour necrosis factor-α
tPA	Tissue plasminogen activator
TRH	Thyrotropin-releasing hormone
tRNA	Transfer RNA
TSH	Thyroid-stimulating hormone
TXA	Thromboxanes (TXA$_2$)
U	Uracil
URE	Upstream regulatory element
UV	Ultra violet
V	Velocity, rate of reaction
V_0	Initial rate of reaction at 0 seconds
V_d	Volume of distribution
VIP	vasoactive intestinal polypeptide
V_{max}	Maximal rate of a reaction
WBC	White blood cell

The scientific basis of therapeutics

ELSIE E. GASKELL

It was a spring Sunday evening and Sarah had just switched the kettle on to make herself a cup of coffee. She had been up all night revising for her exams and had fallen asleep a few hours ago, only to be awakened by the draught banging the kitchen door closed. Her flatmates had left the windows of their student halls accommodation open all day and she was already feeling the annoying symptoms of her hay fever (seasonal allergic rhinitis) settle in. Her eyes were puffy and itchy, her nose was beginning to run, and she had a dull headache developing. She did not need this with her exams looming, especially as she had never got round to going to the pharmacy to pick up some antihistamines. She would now have to catch the bus to town to find an out-of-hours pharmacy and get some cetirizine tablets. She knew she would feel so much better having taken the antihistamine medication but this would definitely set back her revision schedule, and the therapeutics exam was in the morning!

For us to understand how drugs like cetirizine work, and how taking these will alleviate Sarah's hay fever symptoms, we must first explore our own bodies at the molecular and cellular level. Only then will we begin to understand how such compounds may bring about changes in the cellular and physiological processes, resulting in a therapeutic effect and ultimately benefiting our health.

This book aims to unravel some of the fascinating aspects of physiology and pharmacology, and present them in a way that will explain the changes caused by drugs. This first chapter will serve as an introduction to the rest of the book, linking not only the varying topics addressed within this book but other pharmacy-related disciplines covered in the rest of the *Integrated Foundations of Pharmacy* series.

Learning objectives

Having read this chapter you are expected to be able to:

- ➤ Link the different topics presented in later chapters of this book.

- ➤ Relate these topics to how drugs work.
- ➤ Identify the key processes in pharmacokinetics.

1.1 Drugs as therapeutic agents

Under the controlled and stringent conditions within a laboratory, there are many molecules which are capable of eliciting physiological responses when exposed to isolated tissues. These molecules may have the desirable physicochemical attributes that are required for the particular physiological responses to occur, and their pharmacokinetic and pharmacodynamic properties may be adequate. However, not all molecules that are discovered (whether purified from nature or chemically synthesized) and tested in the laboratory are pharmaceutically useful. For a molecule to have a pharmaceutical use it must have an appropriate *therapeutic effect*. Molecules that have relevant therapeutic effects are considered as *drugs*.

> ### KEY POINT
>
> For a drug to have a therapeutic effect it must be able to interact with the body, adjust the processes that are naturally occurring, and consequently cause a change that is clinically beneficial for the patient.

At this point we must also consider molecules produced by our bodies that bring about all the different physiological responses that sustain life. These molecules are endogenous and their concentration, duration of action, and the resulting physiological response are not within our conscious control. Some drugs work in exactly the same way as these endogenous molecules, thus eliciting the same physiological effect, whereas other drugs may in some other way modify the actions of the endogenous molecules. Throughout this book you will encounter a vast number of endogenous molecules, which may have functions varying from conveying a message through the body to initiating an intracellular process. It is crucial we understand these endogenous molecules and the processes they elicit in the body before we look into the therapeutic effects caused by different drugs.

Testing molecules in the laboratory (*in vitro*) helps identify certain characteristics that facilitate their potential therapeutic effect. The effects a drug can induce within a patient are the direct consequence of its chemical and physical interactions with the body. Chemical functionalities present on a molecule will influence its recognition at specific targets and behaviour in different environments. Consider the example of catecholamines, a class of chemicals that all share the same basic core structure, known as the catechol (3,4-dihydroxyphenyl) group. The slight structural differences between some of the members of this class are what make them more selective for different targets. Figure 1.1 shows the chemical differences between three catecholamines and indicates their different pharmacodynamic properties.

 Catecholamines and their cellular targets are covered in more detail in Chapters 5 and 6.

It is vital that as much information as possible about the chemistry of the molecule is discovered before it is tested in living organisms. This is done to minimize any potential toxic effects and improve its activity.

 The *Pharmaceutical Chemistry* book within this series focuses on the relevant aspects of chemistry for pharmacy.

For a pharmacist to dispense or prescribe a molecule as a drug with a known therapeutic action it will have previously been subjected to extensive investigation. The process of drug discovery is long and laborious, involving numerous stages from the non-clinical work in laboratories through to the different stages of clinical trials that ultimately assess the compounds' effectiveness and safety in administration to human patients. This is a costly and stringently regulated process, but with good reason. If a molecule, natural or synthetic, is to be administered as a drug to the patient population its long- and short-term physiological effects need to be known.

FIGURE 1.1 Structural differences of three catecholamines: noradrenaline, phenylephrine, and salbutamol. Their structural variations are the reason these molecules have different pharmacodynamic profiles, evidenced by their selectivity for different targets

Core structure of catecholamines	Molecule	Selectivity for target
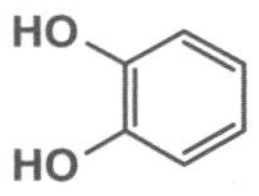 catechol	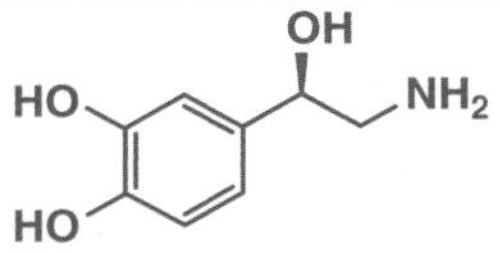 noradrenaline	Non-selective agonist for both α- and β-adrenoceptors
	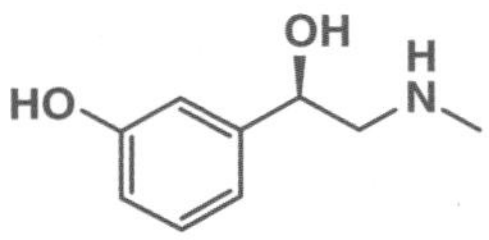 phenylephrine	Selective agonist for α_1-adrenoceptors
	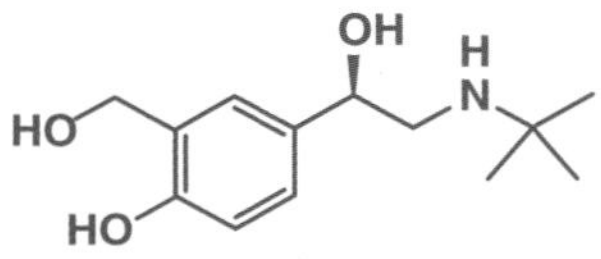 salbutamol	Selective agonist for β_2-adrenoceptors

 You will find more information about dispensing and prescribing, as well as other aspects of the pharmacy profession, in the **Pharmacy Practice** book in this series.

In many cases it is not sufficient to administer the raw drug molecule alone, as it would not necessarily reach the desired location within the body, thus would not cause the required therapeutic effect. This is why drug molecules that can cause a therapeutic effect need be appropriately formulated into *medicines*, and the pharmaceutical properties of the formulation fully understood to enable the desired therapeutic effect.

 Formulation is a section of physical pharmaceutics, some aspects of which are introduced in the **Pharmaceutics** book within this series, along with other topics relevant to drug development.

In this book we will consider the basic concepts of the *scientific basis of therapeutics*, often shortened to *therapeutics*. Therapeutics covers not only the **pharmacology** of a drug but also relies on information from allied disciplines including human physiology, anatomy, pathophysiology, molecular biology, and biochemistry to provide an understanding of the predicted responses the drug would elicit in the body.

KEY POINT

To ensure the best outcome for the patient it is important that pharmacists understand the therapeutic effect of the drugs they are working with, in addition to appreciating the possible interactions and side effects that could arise due to taking the medications.

SELF CHECK 1.1

When is a molecule considered a drug?

1.2 It all starts with the cell—the 'unit of life'

To begin to understand the basic concepts of pharmacology, the effect drugs have on the body (pharmacodynamics), and how the body deals with the administered therapeutic compounds (pharmacokinetics), it is important to understand the physiology of living organisms.

At the cellular level

Cells house the entire code of life, the genetic code, in addition to the required machinery that functions to maintain and perpetuate life at the microscopic level. This is why many consider the cell the 'unit of life'. All the proteins and enzymes we need to build our cells and enable cells to undertake the processes required to sustain life are encoded in the biomacromolecule called deoxyribonucleic acid (DNA), which is the basis of the genetic code. Consequently, whenever there is an irregularity with the genetic code not only will it affect the physiological functioning of cells and organs, ultimately resulting in a clinical condition in the patient, but this outcome can also be passed on from generation to generation. This is why health care professionals will sometimes, as part of a health check, ask for a medical history of the family.

 Section 2.3 'The instruction manual of life: the genetic code', and Section 2.5 'What can happen when things go wrong?' within Chapter 2 of this book will go into more details about the genetic code and the conceptually linked clinical disorders.

SELF CHECK 1.2

Why may it be important for health care professionals to establish the family medical history?

The boundary that separates the inside of the cell from the outside is called the *cellular membrane*, and from a therapeutic point of view this is an important concept. The cellular membrane accommodates a large number of biomacromolecules within its simple lipid bilayer structure and it is with these that the cell regulates the exchange of molecules between its internal and external environment. Some drugs have to interact with the cellular membrane for their therapeutic effect to be manifested. This interaction could trigger a series of subsequent changes within the cell that result in what we observe as the therapeutic effect. To even begin to understand these inter- and intracellular physiological changes that arise owing to the actions of drugs we must first have an appreciation of the cell as a whole.

 Section 2.1 'The unit of life: the cell', within Chapter 2 of this book, will explain in more detail the biology of cells.

KEY POINT

Even though cells are considered the building blocks of tissues, organs, **body systems**, and ultimately living organisms, they are complex structures in their own right.

Energy requirements

Cells regularly undergo a diverse range of processes which sustain life. These processes may well be affected by the presence of a drug molecule. Nonetheless, a cell requires *energy* for these processes, which it must be able to harness from the nutrients available. The energy converted and used by cells is in the form of adenosine-5'-triphosphate (ATP), shown in Figure 1.2. The

FIGURE 1.2 The structure of adenosine-5'-triphosphate (ATP). The red arrows indicate the bonds that are hydrolysed to release phosphate groups. The transfer of these phosphate groups to other molecules results in energy being released from ATP

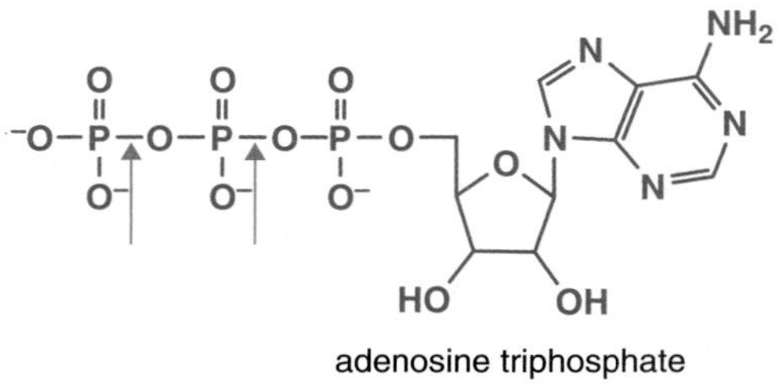

adenosine triphosphate

energy is actually within the phosphate groups of ATP molecules and is released when these high-energy phosphate groups are hydrolysed and transferred to another molecule.

The amount of ATP produced by a cell, thus the *free energy* available to the cell, is dependent on a number of factors, including the type of cell, its physiological state, its external environment, and the nutrients available. Metabolism covers the processes by which cells, and therefore the body, generate and subsequently use energy. It explains why we need to eat food to have enough energy to survive and for our bodies to be able to conduct everyday processes. Box 1.1 looks further into how our body uses energy.

Some nutrients are preferentially metabolized by the body. Any energy that is not used up in the body gets stored in more complex compounds such as fat and glycogen, which are then broken down to release ATP when required. As the metabolism of nutrients lies at the centre of any cellular and bodily function, any abnormality in the sequence of reactions involved will lead to abnormal metabolism and possibly a metabolic disorder. For example, type I tyrosinaemia is a hereditary metabolic disorder involving the catabolism of the amino acid tyrosine. As a consequence, toxic metabolites of tyrosine accumulate in the body, which in turn affect the liver and kidneys as well as causing mental retardation. Therapeutic agents like nitisinone can alleviate the clinical symptoms that arise due to this metabolic disorder by preventing the formation of the toxic metabolites.

 Metabolism and its associated abnormalities are covered in more detail in Section 3.3 'Metabolic pathways and abnormal metabolism', within Chapter 3 of this book.

SELF CHECK 1.3

Why is metabolism central to all cellular processes?

Maintaining an equilibrium within the body

In addition to absorbing nutrients and harnessing energy, cells maintain a balance both within themselves and between surrounding cells. The body has to be able to regulate its internal environment for it to be able to perform its routine physiological functions. This balance is known as maintaining homeostasis. All cellular processes, exchange of nutrients and ions across the cellular membranes, production and use of energy, and so on, are regulated to the highest extent, not only to achieve maximum gain from the resources available but also to ensure the physiological parameters are kept as close to the *set-points* as possible. Feedback mechanisms exist within the body (both positive and negative) to enable it to adjust to the external environment and alter its processes and maintain the physiological parameters within the appropriate limits.

 You can read more about homeostasis in Chapter 8 of this book.

BOX 1.1

Energy usage by our bodies

Our bodies use the energy produced from the food we consume in a number of different ways, which we can broadly divide into two sections:

- Internally—conducting various cellular and bodily processes outside of our conscious thought (for example, the active transport of molecules across the cellular membrane).

- Externally—generating heat and conscious physical activity (for example, running).

Even when we are resting or asleep and do not appear to be physically using any energy our bodies are maintaining vital functions, including heart beat, digestion of food, regulation of fluid levels in cells and tissues etc., all of which require a considerable amount of energy. All individuals have a basal metabolic rate (BMR, measured in calories per unit of time), which identifies the amount of energy that our bodies need to sustain life in a resting state. Everyone may have a different BMR, as it is influenced by genetic as well as environmental and fitness factors. However, when we are awake and going about our daily lives our body's requirement for energy increases. It is not just the physical activities that drain the body's energy resources. Even now, while you are sitting there reading this text your brain is using a considerable amount of energy to maintain the electrical stimulation and facilitate the cognitive processes underway to help you think, learn, and remember.

KEY POINT

The concept of homeostasis is principally simple—keep everything in the body in harmony and maintain a status quo. Nonetheless, the underlying systems that are involved in this 'simple' concept are extremely complex but essential in maintaining health.

Physiological parameters such as body temperature, blood pressure, fluid volumes within the tissues, the composition of blood etc. all have a certain constancy that the body aims to maintain under all circumstances. The circulatory system and blood play important roles in facilitating the maintenance of homeostasis within the body; however, this regulatory role is not their only purpose.

SELF CHECK 1.4

Describe what is meant by homeostasis.

The circulatory system

Blood is a complex fluid consisting of a number of components, all with distinct functions, and on average representing 7–8% of our body weight. It is the body's main transport system—for absorbed nutrients from our gut to the rest of the body, gases (carbon dioxide and oxygen) to and from the lungs, various **messenger signals**, as well as waste products. Therefore, it also has a central role for distributing any systemically administered drug around the body and bringing it to its cellular target.

In addition, blood has an immunological role, providing the required defence mechanisms against invading foreign agents. As such it is vital that it is in full working order. Often, tests assessing the composition of blood samples are used as telltale signs of any changes within the body or malfunctioning tissues or body systems.

KEY POINT

The blood has three main functions: regulation, transport, and protection.

 The blood and its components are covered in more detail in Chapter 9 of this book.

1.3 **Communication is vital**

Maintaining homeostasis within the body as a whole relies on the functioning of virtually every body system, and the different communication systems we have in place (the nervous and endocrine systems) play crucial roles. They enable the exchange of information within the body, facilitating the integrated functioning of cells and tissues.

KEY POINT

In order for a complex organism such as the human body to be able to coordinate its functions, all the different systems must communicate and harmonize their processes.

There are different ways in which cells exchange vital information and thus facilitate an orchestrated response, and this primarily depends on which cells are communicating. Adjacent cells exchange molecules by direct contact of their cell membranes via **gap junctions**. Cells that are further away from each other secrete diffusible endogenous molecules (messenger signals) into the surrounding environment, which initiate specific responses in the recipient cells. This is achieved through the interaction of the messenger signals with different types of proteins, known as **receptors**. Receptors can be embedded in the cellular membrane (for water-soluble molecules) or located intracellularly (for lipid-soluble molecules).

There are different types of receptors and these are discussed in more detail in Section 4.2 'Basic concepts of pharmacodynamics', within Chapter 4 of this book.

SELF CHECK 1.5

What is the role of receptors in facilitating cellular communication?

Long-distance communication

The body's long-distance messenger signals, which have to travel via the blood from one part of the body to the target organs and cells, are called **hormones** and are secreted by specific cells within **endocrine glands**. Examples of endocrine glands include the hypothalamus, pituitary gland, thyroid gland, and pancreas; and examples of some of the hormones they secrete include dopamine, growth hormone, thyroxine, and insulin, respectively. Hormones should not be confused with *autocoids*, also known as local hormones, which do not have to be transported by the blood as they act on neighbouring cells. An example of an autocoid is histamine, which is produced by mast cells in response to local irritation.

 The autocoid and hormonal communication systems in our bodies will be covered in Chapter 7 of this book.

The other body system that has significant importance in facilitating the communication between cells, which may be some distance away, is the nervous system. The nervous system consists of the *central nervous system* (including the brain and the spinal cord) and the *peripheral nervous system* (divided into the *somatic*—voluntary and *autonomic*—involuntary branches). Cells of the nervous system secrete messenger signals called **neurotransmitters**. Signal transmission via the nervous system tends to be faster and more accurate than that produced by the endocrine system. This is because the neurotransmitters are released in close proximity to the target cell, even though the cells communicating may be some distance apart. This is due to the unique structure of nerve cells

(neurons), with their long tentacle-like projections. The longest of these projections is called an *axon* and can be over a metre long. Axons enable an electrical signal to be transmitted from one part of our body to another, where it is involved in releasing neurotransmitters in the vicinity of the target cell. The physiology of the nervous system thus circumvents the circulatory system required for endocrine communication.

 The cellular communication supported by the peripheral nervous system and the physiological responses it facilitates are covered in more depth in Chapters 5 and 6 of this book.

SELF CHECK 1.6

Explain the main differences between hormones and neurotransmitters.

Cellular responses

The molecule acting as a messenger signal (sometimes referred to as a **ligand**) binds to a specific site on a receptor, called the *binding site*. This *activates* the receptor and in turn initiates a cellular response that results in further changes occurring, but now *within* the target cell. These changes may only be transient, lasting less than a millisecond, or can last longer, a few seconds.

KEY POINT

The outcome of receptor activation depends on the concentration of the messenger signal present in the vicinity of the tissue, the type of cell and receptor being activated, and the specific intracellular processes that are initiated, as well as the presence of any feedback mechanisms that may, in turn, 'switch off' this particular process.

Water-soluble (**hydrophilic**) messenger signals, for example the neurotransmitter noradrenaline, which have been secreted in response to an external or internal stimulus, must be able to bind to the target receptors present in the cell membrane in order to convey the information. By contrast, lipid-soluble (**lipophilic**) messenger signals, for example the hormone testosterone, due to their physicochemical properties diffuse through the cellular membrane. Once inside the cell they bind to the intracellular receptors and initiate a cellular process.

Receptors can be very selective for the endogenous messenger signal that activates them. However, a molecule not innately produced by the body, for example a drug, may also activate these receptors and thus bring about the same cellular response, if it has the appropriate physicochemical characteristics required by the receptor.

KEY POINT

A drug which binds to and activates certain receptors can bring about the cellular response required to change the environment of the cell and organ, thus having a beneficial physiological effect for the patient.

There are varying physiological responses possible as a consequence of the action of either the endogenous molecule or the drug, and this primarily depends on the type of receptor activated. For example, skeletal muscles contract due to depolarization of the muscle cell membranes. This depolarization is caused by opening of the nicotinic acetylcholine receptors (nAChR), an ion channel shown in Figure 1.3, which allows the passage of ions through the cellular membrane into the cell. Drugs like suxamethonium chloride mimic the action of the endogenous agonist acetylcholine, and can cause muscle paralysis. Such drugs belong to a class of neuromuscular blocking drugs and are used in conjunction with anaesthetics for facilitating surgical procedures.

Similarly, some drugs can cause relaxation of certain tissues. Drugs such as salbutamol and salmeterol are often used in inhalers for patients with asthma. These compounds mimic the action of the endogenous neurotransmitters and bind to the β_2-adrenoceptor, a type of G-protein-coupled receptor shown in Figure 1.4. These activated receptors stimulate the production of 3′,5′-cyclic adenosine monophosphate (cAMP), an intracellular second messenger. This second messenger then initiates a further reaction cascade, which causes a relaxation of the tissue in our lungs—bronchodilation.

FIGURE 1.3 The nicotinic acetylcholine receptor (nAChR) is a transmembrane receptor that is an ion channel. (A) shows the ion channel closed, not allowing the entrance of sodium ions (Na⁺) into the cell. (B) shows the neurotransmitter acetylcholine activating the receptor, thus opening the ion channel, and Na⁺ entering the cell, which leads to the relevant physiological effect

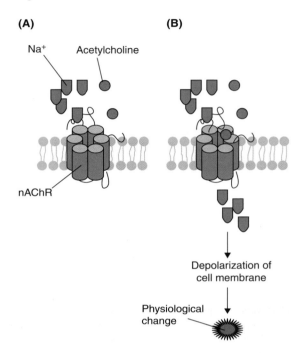

The physiological response caused by an endogenous or drug molecule does not have to be a physical change such as contraction or relaxation of a muscle tissue. The response may simply be a change of concentration of certain molecules. Let's consider insulin, a hormone secreted by the pancreas and involved in glucose metabolism. Insulin binds to the insulin receptor as shown in Figure 1.5. The insulin receptor is a type of tyrosine kinase receptor found in muscle and fat tissues that initiates a series of enzyme-catalysed reactions within the cell. The end result is that there is increased uptake of glucose surrounding the cells. Some diabetes mellitus patients may not produce sufficient insulin to successfully regulate the level of glucose in their blood, thus requiring regular injections of additional insulin to facilitate this whole process.

It is worth noting here that the activation of certain intracellular receptors may cause changes in the expression of specific genes. The thyroid hormone receptor shown in Figure 1.6 is an example of such a receptor.

FIGURE 1.4 β_2-adrenoceptors are G-protein coupled receptors. (A) shows the transmembrane β_2-adrenoceptor associated with the whole G-protein. (B) shows the activation of the adrenoceptor, which stimulates a section of the G-protein (G_α) to bind to the membrane-bound enzyme adenylate cyclase, stimulating the production of cAMP, which triggers the physiological changes

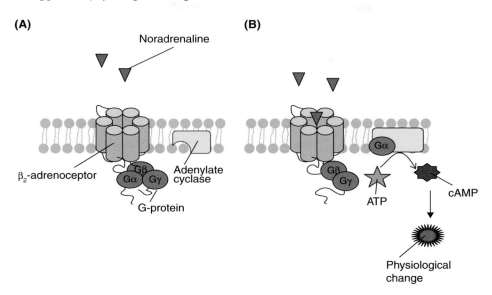

FIGURE 1.5 The insulin receptor is a tyrosine kinase receptor consisting of two identical subunits. (A) shows the insulin receptors not associated with the hormone insulin. (B) shows the association of insulin with its receptors, causing **phosphorylation** of the insulin receptor substrate, autophosphorylation of the receptor, and translocation of the glucose transporter molecules to the cellular membrane, where they transport glucose into the cell

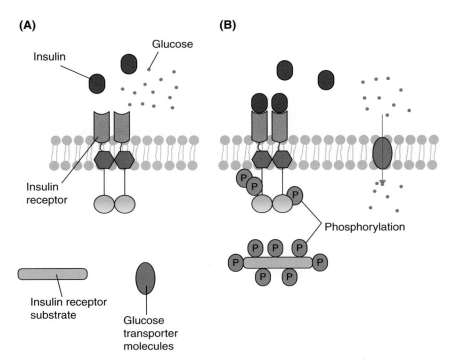

FIGURE 1.6 The receptor for thyroid hormone is intracellular. (A) shows the lipid-soluble thyroid hormone and its receptor on either side of the cellular membrane. (B) shows the diffusion of the hormone through the membrane and its binding to the receptor. This complex can now associate with DNA and regulate the expression of the thyrothropin gene

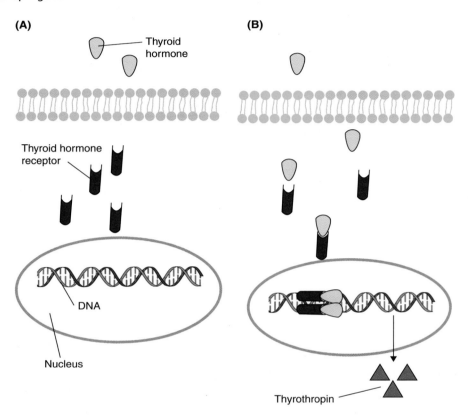

When thyroid hormone binds to the receptor, the complex can associate with DNA and in turn regulate the expression of the gene that codes for another hormone—thyrothropin. Here, the outcome of receptor activation is the alteration of the concentration of a molecule coded for by our genes.

> **KEY POINT**
>
> The varying physiological responses caused by drugs primarily depend on the type of receptor being targeted. Drug molecules can either associate with these receptors or in some way affect the binding of endogenous molecules to the receptors, thus altering the cellular responses.

Cellular responses caused by the activation of *different* receptors present on the *same* cell are not necessar-

ily identical and can in some instances be contradictory. For example, the cells of the sinoatrial (SA) node in your heart have muscarinic receptors (mAChR), which, if activated by the neurotransmitter acetylcholine, cause the heart rate to decrease. The same cells of the sinoatrial (SA) node contain β_1–adrenoceptors, which, if activated by the neurotransmitter noradrenaline, will result in an increase of the heart rate.

> **KEY POINT**
>
> A cell may have a number of different receptors present in its membrane. Therefore, it is able to respond to a range of varying molecules and regulate its physiological outputs to adapt to the external environment.

The distribution of messenger signals and receptors within the body is of paramount importance when considering the physiological changes they bring about. As mentioned above, the neurotransmitter acetylcholine, by activating mAChR, will bring about a reduced cardiac output—a reduced heart rate. However, the same neurotransmitter will induce skeletal muscles to contract by activating nAChR.

Drugs that are able to mimic the action of the endogenous molecules can bring about the same physiological changes, and thus overcome a particular disorder which may exist in the body due to the malfunctioning of the innate processes. If such a drug is administered to a patient, the *distribution* and *selectivity* of the target receptors must be taken into account. Drugs that are intended for therapeutic use are developed with the aim of being as selective as possible for tissues and receptors, thus minimizing undesired side effects. Integration Box 1.1 looks into drug-induced side effects.

> **KEY POINT**
>
> The same messenger signal acting on receptors present in different cell types may induce differing physiological responses.

> **INTEGRATION BOX 1.1**
>
> ## Non-therapeutic effects of drugs
>
> When we talk about the effects of drugs on the body we usually mean the intended beneficial effects. Drug-induced side effects can be described as effects caused in the body that are different to the intended therapeutic effect. These alternative physiological responses caused by the drug may be beneficial or cause harm to the patient. Any undesired side effects are often referred to as adverse drug reactions (ADRs) or adverse side effects. They may come about due to a number of different reasons and are categorized based on their cause as well as severity of the harm evoked in the patient.
>
> Some ADRs are predictable and arise due to the drug interacting with receptors elsewhere in the body and causing undesired physiological effects. Alternatively, they may be a function of the dose of the drug, where increasing the dose causes the ADR, or time, where they arise due to longer exposure of the body to the drug molecule. Other adverse effects may be non-selective and unpredictable because the pharmacological actions of the drug do not explain their occurrence. Unpredictable
>
> ADRs are due to variation in the patient population and may result in reactions like allergies but may also cause **carcinogenic** and **teratogenic** effects.
>
> Adverse effects are considered in more detail in Section 4.3 'Adverse drug reactions, drug interactions, and drug tolerance', within Chapter 4 of this book.
>
> *Pharmacovigilance* is a branch of pharmacology that is concerned with detecting, monitoring, and determining the causes of ADRs, with the aim of preventing them. Information about the undesirable side effects caused by the drug is collected not only during the drug development stages but also after the medicine has been marketed and is taken by a large number of people.
>
> Pharmacovigilance is also considered in Chapter 8 'Conclusion: pharmaceutical care', within the **Pharmacy Practice** book in the series.

1.4 So, how do drugs work?

So far, we have considered the body producing endogenous messenger signals in order to convey information between cells. We have discussed the concept that some drugs are capable of changing cellular and physiological effects by mimicking the endogenous signals. However, there are other modes of action that drug molecules may have, and these are summarized below. Nonetheless, the overarching concept is that drugs, in one way or another, alter the naturally occurring physiological processes, and these changes have positive therapeutic effects.

Drugs as modulators of receptor-induced physiological effects

Any molecule, endogenous or taken as a drug, capable of activating a receptor and thus initiating the associated intracellular physiological effect is called an **agonist**. There are different classes of agonist compounds based on their **affinity** toward the receptors and the **efficacy** of the response they stimulate when the receptor is activated. There are huge numbers of therapeutic compounds that are agonists, mimicking the activity of the endogenous signals produced by the body. For example, the clinically relevant drug sumatriptan is a serotonin receptor agonist. Sumatriptan belongs to a family of compounds collectively known as 'triptans', which are used to alleviate symptoms in patients suffering from migraines.

There are molecules, however, that can bind to a receptor (have affinity for the receptor) but do not activate it (have no efficacy) and thus do not cause any physiological response. These are called **antagonists**. Their association with the receptor either prevents the agonist from binding or prevents the bound agonist inducing the physiological response. There are drug molecules that are receptor antagonists. An example of a clinically relevant antagonist is atenolol, a β_1-adrenoceptor antagonist used for treating hypertension. The medication Sarah was taking for her hay fever, cetirizine, is also an antagonist; it is a histamine H_1-receptor antagonist.

Partial agonists are molecules that have an agonistic effect on the receptor, and thus initiate a response, but only ever achieve a submaximal response. In the presence of a full agonist capable of eliciting a maximal response, partial agonists act like antagonists, competing with the full agonist for the binding site and thus decreasing the overall response possible. Many drugs fall into this category and many side effects can be explained by this concept. Examples of clinically used partial agonists are tamoxifen (see Box 4.3 in Chapter 4) and buspirone, a selective serotonin receptor partial agonist used in the management of anxiety disorders.

KEY POINT

Drugs can modulate physiological responses by stimulating or preventing a particular physiological response from occurring, by interacting with the receptors present within the body.

 Section 4.2 'Basic concepts of pharmacodynamics', within Chapter 4 of this book, will cover the different types of agonists and antagonists in more detail.

Enzymes as drug targets

Receptors are not the only possible targets for therapeutic compounds. We have mentioned above the example of a metabolic disorder, type I tyrosinaemia, and the fact that drugs exist which can alleviate the clinical consequences of this disorder. Nitisinone is an *enzyme inhibitor*, and as such prevents the specific enzyme (4-hydroxyphenylpyruvate oxidase) from catalysing the reaction that produces toxic products. As no toxic metabolites are formed, the clinical outcomes of the disorder are alleviated.

KEY POINT

There are numerous clinically relevant enzyme inhibitors, which function to selectively inhibit innate enzyme-catalysed processes.

If the inhibited enzyme reaction is part of a cascade of reactions this whole process may well be halted, having notable effects on the cell, tissue, and ultimately the body. The complex formation between the enzyme and its inhibitor may be a reversible process, whereby the enzyme is eventually released intact and thus capable of further catalysing the reaction. This is the case for the vast majority of clinically used enzyme inhibitors. Irreversible binding of inhibitors to the enzymes is a less desirable property of some therapeutic compounds.

 Section 3.2 'Enzymes and enzyme inhibition', within Chapter 3 of this book, explores enzymes and enzyme inhibitors in more detail.

It is relevant to mention here that enzymes themselves may be given as therapeutic agents. Patients in whom a certain enzyme is either absent or non-functioning can be given *enzyme replacement therapy (ERT)*, which involves the delivery of the missing or deficient enzyme, usually by intravenous infusion. This overcomes the biochemical barrier that causes the clinical symptoms. An example of where ERT has clinical significance is in the treatment of Gaucher's disease, a hereditary metabolic disorder affecting the storage of fat in certain cells and tissues.

Additionally, enzyme substrates can be administered if the cause of a disease is suboptimal amounts of a particular endogenous molecule. Here, the precursor of that molecule can be given as a drug, which is subsequently converted to the active form by the endogenous enzyme. Levodopa, for example, is used in the treatment of Parkinson's disease and is the precursor of the neurotransmitter dopamine, which is present in low concentrations in patients with this disorder.

Some other modes of drug action

Transport proteins are cell membrane-bound proteins whose role is to transport endogenous molecules across the membranes. They may be considered as receptors themselves, responding to a specific signal by facilitating a cellular process; however, transport proteins are not linked to a sequence of intracellular transformations like those described previously. Generally, they facilitate the uptake of endogenous molecules from outside the cell, resulting in their reduced contact and interaction with the relevant receptors on the target cell. Therefore, drug molecules that bind to transport proteins and inhibit this process can have a therapeutic effect as they prolong the presence of the endogenous molecule within the vicinity of its receptors. These molecules are called *reuptake inhibitors*. An example is methylphenidate, a dopamine and noradrenaline reuptake inhibitor used to treat attention deficit hyperactivity disorder (ADHD). Also, antidepressants like fluoxetine are selective serotonin reuptake inhibitors (SSRIs).

So far we have only considered cellular targets (membrane-bound or intracellular); however, some disorders can be treated by compounds not requiring interaction with any of the cellular components. These compounds alter the surrounding tissue environment. For example, antacids neutralize stomach acid that has refluxed, causing heartburn, and thus alleviate the symptoms. Osmotic laxatives increase the volume of fluid in the large bowel, facilitating the relief from constipation. In both examples there is no direct effect of the therapeutic compound on any of the cellular targets previously mentioned.

Many disorders have genetic causes, meaning the clinical symptoms that are present are fundamentally due to alterations in the genetic code. Some genetic disorders have already been mentioned and the drugs indicated as treatments for these are effective in alleviating the symptoms caused by the disorder. However, they do not deal with the genetic basis of the disorder itself. *Gene therapy* aims to eradicate some genetic disorders by providing an alternative DNA sequence the body can use to produce the relevant protein product. In effect, in gene therapy DNA sequences are used as drugs.

 Section 2.5 'What can happen when things go wrong?', within Chapter 2 of this book, explores the current and future genetic therapies.

Finally, a disease state within our bodies can be brought about by non-host cells, such as viruses, bacteria, fungi, and protozoa. Once they have overcome the natural barriers that exist within a human body, pathogenic organisms may cause harm in a number of different ways, depending on the type of infection, its duration, and the part of the body affected. Most infections in our bodies elicit an immune response, causing the common symptoms associated with infection: fever, redness, swelling, and pain. Some infections are transient and easily eradicated but others are more dangerous and harder to cure. There are different types of drugs used to treat infections and prevent further infection, all with the aim of eliminating the infecting organism and consequently reducing its detrimental effects. This can be achieved by selectively targeting the infecting organism, and thus the therapeutic effect is achieved indirectly by removing the cause of the disease. In these situations drugs can also be taken to alleviate the symptoms associated with infection, and these assist the body's natural defence mechanisms.

SELF CHECK 1.7

Describe five different ways drugs may cause therapeutic effects.

1.5 The effect our bodies have on administered drugs

Irrespective of the details of a specific drug's mode of action, for it to have a given effect on the physiological processes within the body it has to be present within the body and, more specifically, it has to be in close proximity to its target, whether this is a receptor, enzyme or infecting organism. How the drug gets into and is subsequently distributed throughout the body, in addition to how the body handles the drug, are complex processes addressed by *pharmacokinetics*. The science of pharmacokinetics is often divided into four sections: *absorption, distribution, metabolism,* and *elimination,* collectively referred to by the acronym *ADME.*

KEY POINT

This section is intended only as an overview of pharmacokinetics and by no means is all-inclusive. However, it is important for you, at this stage of your study, to appreciate the basis of these concepts.

A is for absorption

There are a number of different routes via which drugs can be administered to the body, and this can be considered the most important step in their use for the treatment of particular disorders. Without appropriate administration, the subsequent absorption of a drug may lead to suboptimal concentrations and distribution throughout the body, resulting in no or inappropriate therapeutic responses. The common routes of administration include: oral, sublingual (under the tongue), inhalation, rectal, topical (directly onto the skin or exposed mucosal areas), and parenteral (injection). They all have different effects on the absorption of the drug and, consequently, its concentration within the body. Figure 1.7 summarizes the common drug administration routes.

 Routes of administration are considered in detail in Chapter 13 'Drug development and delivery', within the **Pharmaceutics** book in the series.

The term *bioavailability* is used to describe the fraction of the dose of drug administered that is present within the body and thus available to facilitate the desired physiological changes. The bioavailability of a drug is linked to the route of administration, as it is dependent on which epithelial tissues are involved in the absorption process. For example, for drugs administered intravenously (directly into the blood), the bioavailability is 100%. However, for a drug administered orally, the bioavailability is greatly reduced as the drug first has to be absorbed through the epithelial layers of the gastrointestinal tract into the circulatory system, which passes through the liver, a major site for drug metabolism and elimination (see Figure 1.7). Drug metabolism is described under 'M is for metabolism' in this section. Therefore, if the same drug can be administered both intravenously and orally, the dose given via the two routes will vary accordingly.

FIGURE 1.7 A schematic diagram of the circulatory system, with the common routes of drug administration indicated. The direction of blood circulation affects the absorption, concentration, and distribution of drug in the body

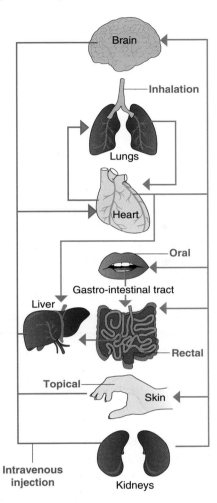

KEY POINT

Drug absorption depends on the physical and chemical characteristics of the molecule, including the water solubility and ionic state (only uncharged molecules can be absorbed across the epithelial membranes).

 The physical and chemical characteristics of drugs are considered in more detail in the *Pharmaceutics* and *Pharmaceutical Chemistry* books in the series.

Most drugs require absorption by the surrounding epithelium and to be present in the blood plasma for them to be distributed effectively. Obviously, drugs which do not interact with any cellular targets, for example the antacids mentioned above, do not need to be absorbed for them to have a therapeutic effect.

D is for distribution

Once in the circulatory system the drug is distributed to different parts of the body, including the fluid surrounding the tissues (interstitial fluid), the intracellular fluid (cytoplasm) and the adipose tissue. However, where it accumulates depends on the physicochemical properties of the specific molecule.

The more lipophilic the drug molecule, the more it will be sequestered from the blood into the adipose tissue; more hydrophilic molecules will remain dissolved in the blood plasma and interstitial fluid. The *volume of distribution* (V_d), often considered in pharmacokinetic studies, refers to the amount of drug present in the blood in relation to the total amount of drug administered. In other words, it measures the extent to which the drug is distributed in the surrounding tissues. Drugs that are not readily partitioned into the surrounding tissues (for example, the anticoagulant warfarin) therefore have a lower V_d value, closely resembling the volume of blood we have (warfarin has a V_d of 8 litres). Drugs that readily bind to surrounding tissues (for example digoxin, used in the management of atrial fibrillation) or are highly lipophilic (for example, the antimalarial drug chloroquine) have high volumes of distribution (digoxin V_d = 490 litres and chloroquine V_d > 13 000 litres).

M is for metabolism

The major site of drug metabolism is the liver (*hepatic metabolism*). Here, a collection of enzymes, mainly belonging to the *cytochrome P_{450}* family, chemically alter the therapeutic molecule, resulting in drug metabolites with altered physicochemical characteristics, making them easier to eliminate.

Our blood regularly passes through our liver. Any free drug present in the plasma will therefore be exposed to the metabolizing liver enzymes. However, not all the drug present in the blood is readily metabolized as the blood passes through the liver. Drugs can bind to plasma proteins (albumin, lipoproteins, and

α_1-acid glycoproteins) and as such are protected from the hepatic metabolism. This protein binding process is reversible and the amount of freely available drug is in equilibrium with the protein-bound drug. Some drugs bind to proteins more readily than others. This not only affects their eventual metabolism and subsequent elimination but also influences their ability to interact with their targets, as only free (unbound) drug can cause a physiological change.

> **KEY POINT**
>
> Most of the therapeutic compounds administered to the body are eventually metabolized, though some are eliminated from the body unchanged.

> **SELF CHECK 1.8**
>
> Why is the bioavailability of a drug taken orally lower than when it is administered parenterally?

E is for elimination

Elimination of the therapeutic compound starts the moment it is introduced to the body, and is an irreversible process of removing the drug, or its metabolites, from the body.

Excretion of administered drugs is achieved through the kidneys (via urine), the hepatobiliary system (via bile), the intestinal transit (via faeces), the lungs (for volatile compounds), in addition to secretions such as sweat and milk. The kidneys play an important role in passively filtering and actively taking up any unbound drug from the bloodstream. They are also involved in passively reabsorbing drugs back into the bloodstream, prolonging their circulation through the body.

Total *clearance* (*Cl*) of a drug is regarded as the volume of blood that is cleared of drug per unit time, through all the routes available. Clearance varies considerably among patients as it is affected by age, disorders (primarily affecting the liver and kidneys), and some concomitant medication, as well as genetic predisposition (**pharmacogenomics**). It is inversely proportional to the *half-life* ($t_{1/2}$) of the drug. The half-life is used to determine the duration of action of the drug, and simply indicates the time it takes for the plasma concentration of the drug to decrease by half. It is affected by the V_d, as the larger the fraction of drug

not in the plasma (the larger the V_d), the longer it will take to reduce its concentration.

Therapeutic window

Each drug has a *therapeutic window*, which is the range of doses within which there is sufficient drug to cause physiological change without an increased risk of side effects arising or toxic effects occurring in the patient.

The pharmacokinetic parameters covered above: V_d, *Cl*, and $t_{1/2}$, are all important when considering the dose and the **dosage regimen** for the patient. As the dosage regimen of a drug is affected by the clearance of the drug, it may require adjustment in individual patients in order to achieve a constant concentration within the therapeutic window. Figure 1.8 shows the effect administration regimen has on the pharmacokinetic profile of a drug.

Sometimes, a single *loading dose* of the drug, much higher than normal, is administered in order to achieve a blood concentration within the therapeutic window in a relatively short period of time. Any subsequent administration of the drug depends on the dosage formulation used (tablet, intravenous injection, etc.) and the route of administration, but aims to keep the drug's plasma concentration within the therapeutic window. Concentrations higher than the therapeutic window can cause adverse effects, though not all side effects are linked to concentration (see Integration Box 1.1). Adverse side effects arise when the administered drugs have unwanted interactions with cellular targets, or when the drugs themselves, or their metabolites, are toxic to the body.

 Adverse effects and related topics are covered in Section 4.3 'Adverse drug reactions, drug interactions, and drug tolerance', within Chapter 4 of this book.

> **KEY POINT**
>
> To understand any unwanted effects, and consequently avoid them, not only should a drug's mode of action be known but the effect our body has on the drug should be considered too.

> **SELF CHECK 1.9**
>
> What are the dangers associated with the administration of a drug with a narrow therapeutic window?

FIGURE 1.8 (A) The pharmacokinetic profile of a single administration of a drug (indicated by the red arrow). The concentration of the drug in the blood reaches a peak shortly after administration. The concentration then gradually decreases and eventually becomes too low to elicit a physiological response. (B) The pharmacokinetic profile of the same drug, re-administered after set periods of time (indicated by the red arrows). The concentration within the blood is continuously within the therapeutic window, resulting in a constant therapeutic effect

(A)

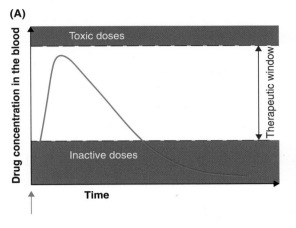

(B)

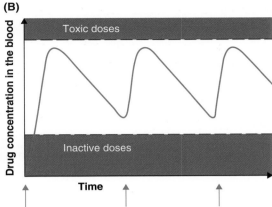

CHAPTER SUMMARY

➤ Pharmacy is complex and multidisciplinary, interlinking an array of different pharmaceutical sciences and practice-oriented disciplines. A sound knowledge base is core to the success of any pharmacy student and future pharmacist. This book provides an integrated approach to the science of therapeutics, but not without continuous reflection and further integration with its associated disciplines: pharmaceutical chemistry, physical pharmaceutics, and pharmacy practice.

FURTHER READING

Boarder, M., Newby, D., and Navti, P. *Pharmacology for Pharmacy and the Health Sciences: A Patient-Centred Approach.* Oxford University Press, 2010.

This pharmacology textbook outlines the subject through an array of clinical disorders.

Pocock, G. and Richards, C.D. *The Human Body: An Introduction for the Biomedical and Health Sciences.* Oxford University Press, 2009.

This textbook covers the anatomy and histology of the human body and includes original research images.

Rang, H.P., Dale, M.M., Ritter, J.M., Flower, R.J., and Henderson, G. *Rang and Dale's Pharmacology.* 7th edn. Churchill Livingstone, 2012.

This is an all-encompassing pharmacology textbook with extensive detail in all aspects of the science.

2

Molecular cell biology

ANDREW EVANS

Our current knowledge barely scratches the surface of the biological processes that occur within our cells at the molecular level. This chapter will, in various ways, underpin the rest of this book by explaining some of the cellular processes that are often altered by therapeutic agents. We will look at the different evolutionary classes of cells that form the basis of all life. In addition, the genetic code will be explored, covering how it is stored, used, protected, repaired, and passed on from generation to generation. Diseases that can come about when our genetic code is damaged and not repaired will also be discussed. Even the little we do know about molecular cell biology never fails to amaze me, and after reading this chapter I hope you will be amazed too!

Learning objectives

Having read this chapter you are expected to be able to:

➤ Appreciate the differences between prokaryotic and eukaryotic cells.

➤ Explain the basic structure of nucleotides.

➤ Identify the essentials of DNA structure and how DNA is packaged into the cell.

➤ Know how the genetic code works.

➤ Understand how the code is transcribed and translated into protein.

➤ Describe the basic processes of DNA replication.

➤ Understand some basic mechanisms the cell uses to repair DNA damage.

➤ Explain how, as a consequence of mutations, disease might arise.

➤ Recognize the potential of current and future genetic therapies.

2.1 The unit of life: the cell

We can consider the cell as a self-sustained parcel or *unit of life*. This is because cells are present in all living organisms, whether they exist as a single cell or a collection of cells. Cells also contain all the necessary information and machinery to be able to sustain and perpetuate life.

Common features of cells

Despite the vast diversity of life we have on Earth, living organisms are made from cells and all cells have some common features. These include:

- They have all inherited from their predecessors, and can also pass on to their **daughter cells**, the *instruction manual of life*, or the **genetic code**. We will look into this further in Section 2.2 'The molecules of life: basic structure of DNA and RNA'.

- All cells also have a *cellular membrane*, sometimes called a *plasma membrane*. This cellular membrane is a **lipid** bilayer (consists of two layers of lipids) that acts as a barrier to separate the inside of the cell from the external environment. Although some molecules are able to pass though the cellular membrane, it is a selective barrier to most. Many **proteins** are incorporated into the cellular membrane, and these have various functions. Some of them act as transporters or channels that use energy to actively transport specific molecules such as nutrients, waste products, or ions into or out of the cell. Others form selective transporters or channels that allow passive (energy-free) movement of specific molecules across this barrier. Thus, the cellular membrane regulates what enters and exits the cell.

- All cells carry out **metabolism** to produce energy and make new molecules.

- Finally, all cells are capable of movement of some form. This can be just the movement of individual components within the cell to assist certain cellular processes, or the movement of the whole cell.

 Section 3.3 'Metabolic pathways and abnormal metabolism', within Chapter 3, will help you understand metabolism in more detail.

Basic cell structure of prokaryotic and eukaryotic cells

Most cells can be classified into one of the two main evolutionary branches. The first branch is the *prokaryotes,* which are the simplest forms of life and tend to exist as single-celled entities. The second branch is the *eukaryotes*; these cells are more complex and whereas some can exist in a single-celled form, others form multicellular organisms like you and me. Although there are similarities between all cells, there are also fundamental differences between these two classifications, which we will look at next.

The fundamental difference between a prokaryotic and eukaryotic cell is the fact that eukaryotes have a membrane-bound **nucleus** (also known as a *karyon*) that separately contains the genetic material, whereas prokaryotes do not. Prokaryotes are relatively small cells, with a total volume about a thousand times smaller than that of a typical eukaryotic cell. We have about ten times more prokaryotic cells in our gut than the eukaryotic cells that make up our whole body!

Eukaryotes are subdivided into **plants**, **animals**, **fungi** or **protists** (protozoa and algae), whereas prokaryotes have traditionally been subdivided into *bacteria* and *archaea*. You are probably familiar with bacterial **genera** such as *Streptococcus, Salmonella,* and *Escherichia* because they are associated with disease. However, the genera of archaea, for example *Haloarchaea* or *Pyrococcus*, are not that well known, probably because they are generally not thought to cause disease. Archaea are about the same size and shape as bacteria, but the cell wall and cellular membrane of archaea are made of different components to those of bacteria. Box 2.1 further explores the classification of archaea.

Surrounding the cell

All cells have a cellular membrane, which allows the passage of water and gases such as oxygen and carbon dioxide. Therefore, if there is a higher **osmotic pressure** within the cells than in the environment outside

BOX 2.1

Bacteria versus archaea

Archaea are often found living in some of the most extreme environments found on Earth, in conditions where humans would not be able to survive. These include high temperatures, often above 100°C, such as those seen in deep-sea vents. Some of the chemical processes carried out by archaea are similar to those of bacteria; however, some have also been found to be similar to those of eukaryotes. More recently, a number of genetic differences and similarities between archaea, bacteria and eukaryotes have been revealed. Because of these differences and similarities, many researchers now believe that archaea should actually be classified separately from prokaryotes and eukaryotes.

the cells, water can enter, swell, and ultimately burst (*lyse*) the cells. This would result in cell death (*cytolysis*). If you look at Figure 2.1(A) you will see that prokaryotic cells, in addition to the cellular membrane, usually also have a rigid *cell wall* that is sometimes surrounded by a *capsule*. The rigid cell wall is a simple yet effective way of preventing cytolysis from occurring. It consists of *peptidoglycans*, chains of amino acids and sugars which have a strong lattice structure. The cell wall is the target of a number of classes of antibiotic drugs that we use, including the β-lactam antibiotics (such as penicillin), that work by inhibiting the production of the cell wall. The capsule in prokaryotic cells provides an additional protective layer. It is a slimy layer, full of polysaccharides and water, which acts as a shield preventing the cell from being destroyed by other cells, or being dehydrated, and it can even act as a reserve of nutrients for the cell.

Plants, fungi, and algae also have a peptidoglycan cell wall, whereas animal cells do not because they are able to control their osmotic pressure. Animal cells, however, have a *cytoskeleton*, which allows them to adapt to a variety of shapes and structures. For example, a single nerve cell in our body can be extremely thin but over 1.5 metres in length, whereas cuboidal epithelial cells are an approximately 0.02 millimetre cube! Figure 2.1(B) shows a typical eukaryotic animal cell.

You will also notice from Figure 2.1 that prokaryotes often have structures protruding from their surface whereas eukaryotes generally do not. The *pili* are hair-like additions that allow the cell to attach to other cells and facilitate the sharing of genetic material. Some prokaryotes also have tail-like features called *flagella*, which rotate to propel the cell forward, enabling the cell to swim.

Inside the cell

The interior of all cells (prokaryotic or eukaryotic) is bathed in a highly regulated aqueous solution called the *cytoplasm*. Within the cytoplasm of eukaryotes

FIGURE 2.1   Basic structure of (A) prokaryotic and (B) eukaryotic cells

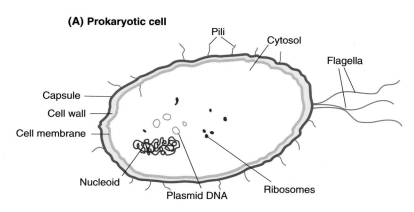

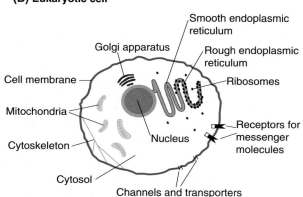

there are membrane-bound structures, collectively known as **organelles**, each filled with its own specific fluid. An important organelle already mentioned is the nucleus. Prokaryotes do not have this level of structure. As they have no nucleus, prokaryotes usually carry their genetic code in the form of a condensed single circular DNA molecule within the cytoplasm. This is sometimes referred to as the *nucleoid*. In addition, prokaryotes can also have relatively small circular DNA molecules called *plasmids*. Plasmids can be transferred from one cell to another via their pili. This means that certain genes can be shared. For example, if a bacterium develops genetic resistance to an antibiotic it can quickly share this with the whole population, making it resistant to that antibiotic.

The organelles present in eukaryotic cells carry out specialized functions within the cell, similar to the way our organs carry out specialized functions in our bodies. Different eukaryotes have defined differences in their organelles, some of which are described below:

- *Chloroplasts* are only present in plants and protists. These organelles can generate energy from sunlight.
- *Mitochondria* are present in virtually all eukaryotic cells. They are often described as the power stations of the cell because they produce most of the cell's energy in the form of **adenosine-5′-triphosphate (ATP)**. Mitochondria are about the size of a typical prokaryotic cell, and eukaryotic cells may contain in excess of two thousand of them. They have their own DNA, the origins of which are discussed in Box 2.2.
- *Ribosomes* are where proteins are assembled from amino acids.
- The *endoplasmic reticulum* is a network of membrane-bound folds that surrounds the nucleus. The rough endoplasmic reticulum appears rough because it is covered with ribosomes, which carry out the processing and folding of new proteins. Lipids are produced by the smooth endoplasmic reticulum, which does not contain ribosomes.
- The *Golgi apparatus* further modifies and packages the proteins and lipids produced by the endoplasmic reticulum.

BOX 2.2

Endosymbiotic theory

The endosymbiotic theory was first proposed in 1905 (*endo* means 'within' and *symbiotic* means 'cooperative'). Owing to some striking similarities between free-living prokaryotes called cyanobacteria and the chloroplasts found in plant cells, it was proposed that chloroplasts were once also free-living cyanobacteria-like prokaryotes. However, it was not until the 1960s, with the use of more advanced techniques in cell and molecular biology, that this theory was taken seriously. Many scientists now believe that chloroplasts and mitochondria are actually direct descendants of prokaryotic cells that were engulfed by ancestral eukaryotes, and existed and evolved within them! These organelles have their own DNA, which has similarities to prokaryotic DNA, and they replicate independently of the cell.

 You will learn about the process of ATP production in Section 3.3 'Metabolic pathways and abnormal metabolism', in Chapter 3.

From single cells to multicellular organisms

Eukaryotic cells often exist within multicellular organisms, like the human body, and communication between cells is essential for the survival of the whole organism. Cells communicate with each other by releasing or receiving messenger molecules. These messages are received by **receptors** found within cells or on the cellular membranes of responding cells.

 Read more about communication between cells and tissues in Chapters 5, 6, and 7. See Chapter 4 for more details on receptors.

In addition to communicating with each other, cells within multicellular organisms must become *specialized* to fulfil specific needs required by the organism as a whole. These specialized cells make up **tissues**, organs, and body systems. Specialization of cells occurs during development, by a process called **differentiation**. As cells divide during development, the daughter cells receive all the genetic information from the parental cell.

However, the genes expressed by the daughter cell may be different compared with those of the parental cell, making the daughter cell functionally different. This concept is further explained in Box 2.4 later in the chapter. Clearly, the expression of genes has to be a highly controlled process, with only the appropriate genes being used at the appropriate time in the appropriate cells. In cancer, cells often lose this control and express a number of genes inappropriately, which can lead to the progression of the cancer. Before we explore gene expression, however, we must first draw our attention to the molecules that make up the genetic information.

KEY POINT

All our cells contain identical genetic information; however the cells differentiate to make up the various types of tissues, organs, and body systems that we are made of.

SELF CHECK 2.1

What are the two major physical differences between prokaryotes and eukaryotes?

2.2 The molecules of life: basic structure of DNA and RNA

So far, we have covered the basic structure of the unit of life, the cell, and now we will look at the basic structure of what may be considered the molecules of life—deoxyribonucleic acid (DNA) and ribonucleic acid (RNA).

Bases, nucleotides, and the phosphodiester bond

You can think of the *genetic code* as a language. As with most languages, the genetic code needs letters to make up words. The letters come in the structural form of nucleotides, consisting of three distinct parts:

1. *Heterocyclic bases*. There are four different nucleotide bases: *cytosine* (C) and *thymine* (T), which have a pyrimidine structure; and *adenine* (A) and *guanine* (G), which have a purine structure. In RNA molecules, thymine is replaced by a structurally similar pyrimidine, *uracil* (U).
2. *Sugar molecules*. These are covalently bonded to the cyclic bases. Two different five-membered sugars are used: ribose in RNA and 2′-deoxyribose in DNA. The heterocyclic base and the sugar together make up a *nucleoside*.
3. *Phosphate groups*. The addition of one or more phosphate groups to the nucleoside is needed to form a nucleotide. These phosphate groups are covalently bonded to the sugar molecule.

Figure 2.2 shows the basic structure of a typical nucleotide. Nucleotides are not only found within DNA and RNA molecules, as discussed in Box 2.3.

Chapter 9 'Chemistry of biologically important molecules', in the **Pharmaceutical Chemistry** book within this series, will explain the chemistry of the nucleotides in more detail.

FIGURE 2.2 Basic structure of a nucleotide

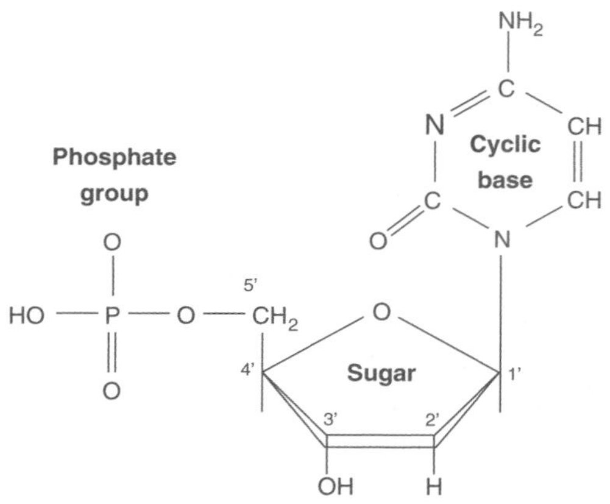

BOX 2.3

Nucleotides have other functions

While monophosphate nucleotides are used in DNA and RNA molecules, variants of nucleotide structures are used throughout the cell for various other functions. For example, adenosine-5′-triphosphate (ATP), produced by the mitochondria to store energy, is in fact a nucleotide comprised of the base adenine attached to a sugar molecule (ribose) that has a chain of three phosphate groups attached. **3′,5′-Cyclic adenosine monophosphate (cAMP)**, a messenger molecule within cells, is another variant. Here, the phosphate group is attached to two different places on the ribose molecule, forming a ring structure.

In DNA and RNA, the nucleotides are joined together into a strand or chain by the nucleotide phosphate groups—forming a covalent bond called the *phosphodiester bond*. The carbon atoms within the nucleotide sugar unit (ribose or 2′-deoxyribose) are conventionally numbered by chemists as 1′, 2′, 3′, 4′, and 5′ (one to five prime), according to their position in the sugar ring structure, as shown in Figure 2.2. The phosphodiester bond forms between the 3′ carbon of the sugar molecule in one nucleotide and the 5′ carbon of the sugar molecule in the next nucleotide in the chain. This nucleotide bonding is continued to form the large chains of DNA or RNA, as shown in Figure 2.3. The nucleotides are always joined in the same direction, 5′ to 3′, with the bases protruding out from the *sugar–phosphate backbone*.

KEY POINT

DNA and RNA are made up of nucleotides that are joined together by phosphodiester bonds.

When reporting the sequence of nucleotides within a DNA or RNA molecule we usually only report the letters of the cyclic bases (G, C, A, and T or U). There is a convention that sequences of nucleotides are always written in the 5′ to 3′ direction; for example, 5′-ATTACGAGT-3′.

FIGURE 2.3 **The nucleotide strand**

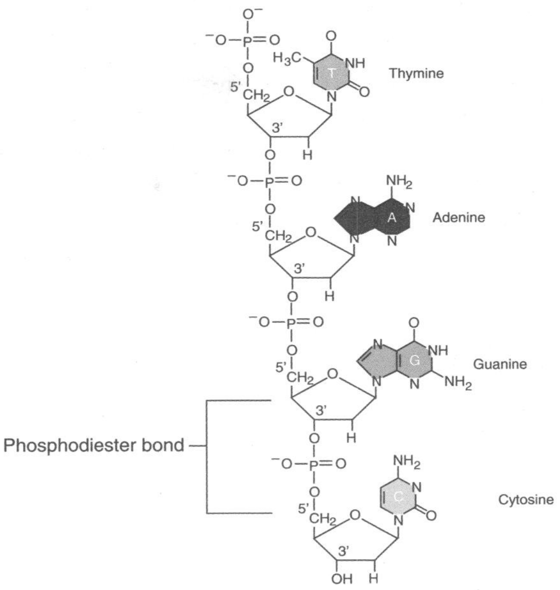

KEY POINT

As with words in the English language, the genetic code only makes sense if the bases are read in sequence and in one direction, from 5′ to 3′.

The DNA double helix

Normally within our cells, DNA is in a double-stranded form. Here, the two separate strands of DNA are joined together in an *antiparallel* fashion, with one strand running in the 5′ → 3′ direction and the other running in the opposite, 3′ → 5′ direction. These strands are joined together by hydrogen bonds between their bases to form *DNA base pairs*. Guanine (G) and cytosine (C) exclusively bond via three hydrogen bonds to each other, and adenine (A) and thymine (T) exclusively bond together via two hydrogen bonds. Because of this, the base sequence of each strand is matched, and so the two strands are said to be *complementary*. Because the complementary strands are joined at their bases, the bases are on the inside and the sugar–phosphate backbone from each strand is on the outside of a double-stranded DNA molecule. This arrangement is illustrated in Figure 2.4(A).

FIGURE 2.4 The structure of double-stranded DNA. (A) Complementary base pairing. (B) The right-handed DNA double helix

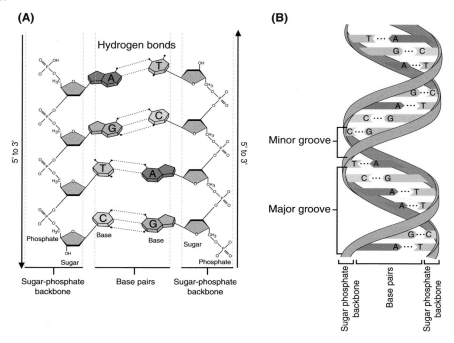

The double-stranded DNA actually occurs as a right-handed *double helix*. It might help to imagine the double helix as a twisted ladder, with the rungs as the base pairs stacked up on each other. The handrails on each side of the ladder are the two sugar–phosphate backbones. Each rung of this imaginary twisted ladder is rotated around the central axis by about 36° relative to the rung beneath. Therefore, the double helix makes a complete turn about every 10th base pair. The DNA double helix is a *negatively charged* molecule because of the negative charge of the phosphate groups on the outside of the molecule. The spaces between the two sugar–phosphate backbones in the helix give rise to two grooves, known as the *minor groove* and the *major groove*. Figure 2.4(B) shows the structure of the DNA double helix.

The DNA double helix is a very stable structure and this helps to protect our genetic code. Single-stranded DNA or RNA molecules are relatively easily damaged. The bases that hold the genetic code are protected on the inside of the double helix. As the double helix has complementary strands of DNA, if one strand is damaged there is always a complementary strand for reference during repair. In this way, the very structure of the double helix helps to maintain the integrity of our genetic code.

SELF CHECK 2.2

This is the sequence of a section of a DNA strand: 5′-ATTACGAGT-3′. What is the complementary sequence of the other strand?

SELF CHECK 2.3

List the three main differences between DNA and RNA molecules.

Higher order genetic structures

Our cells are virtually too small to see with the naked eye, yet the double-stranded DNA contained within the nucleus of just one cell, if measured from end to end, would be about 2 metres in length. It has been estimated that if all of the double-stranded DNA in all the cells in our body was lined up end to end, it would reach to the sun and back about seven times! Clearly, there has to be some form of highly organized packaging in order to fit our DNA into a cell's microscopically sized nucleus.

The DNA molecules we have in each cell are very long. They are, however, extremely thin, measuring only about 2 nanometres (nm) in diameter (that is

2 millionths of a millimetre). The nucleus is the largest organelle in our cells. Still, it requires a lot of organized tidying away for a cell to fit all its genetic material into its nucleus and for it to know exactly where each bit of information is located.

Our genetic material is organized structurally in three levels. The *primary structure* is the DNA nucleotide strand and the *secondary structure* is the formation of the DNA double helix. Both of these levels we have already discussed. The third level, the *tertiary structure*, is when the double helix is condensed into structures that can fit into a nucleus.

To package DNA into a nucleus, the double-stranded DNA initially forms a complex with positively charged proteins called *histones*. There are different types of histone: H2A, H2B, H3, and H4 are the 'core' histones, and H1 is the much larger 'linker' histone. Two of each of the core histones form protein complexes known as histone octamers. These act as spools for DNA and are central to DNA packaging. The double-stranded DNA wraps around a histone octamer forming a *nucleosome*. Because the histones are positively charged, they form ionic bonds with the negatively charged phosphate groups of the DNA wrapped around them. This holds the nucleosome together. Histone H1 binds to the DNA at the point where the DNA strand enters and leaves the nucleosome. This draws the nucleosomes closer together. The resulting 'string' of nucleosomes is known as the *10 nm fibre*. This 10 nm fibre is folded in a 'zigzag' rotational fashion into a structure called the *30 nm fibre*, thus further packaging the DNA. For each rotational turn in this 30 nm fibre there are over a thousand base pairs of DNA.

Another important component in the DNA packaging process is the *nuclear matrix*. This is a linear protein complex that acts as a central scaffold or framework around which the 30 nm fibre is wound further, forming *loops* of fibre that project out from the nuclear matrix. There

are thought to be 18 of these loops for every turn the fibre makes around the central axis of the scaffold. Each of these loops is further coiled or folded to condense the DNA yet further, although exactly how this occurs is at present subject to speculation. In humans, the entire single 30 nm DNA fibre that is wrapped around the nuclear matrix can contain as many as 247 million base pairs of DNA and have over four thousand genes. This whole structure, that is the highly packaged double-stranded DNA molecule, the associated proteins, and nuclear matrix scaffold, together is known as a *chromosome*. Figure 2.5 illustrates the different levels of DNA packaging in a chromosome.

> **KEY POINT**
>
> DNA is super-coiled to form nucleosomes → 10 nm fibre → 30 nm fibre → loops → chromosomes.

The number of chromosomes within the nucleus of eukaryotic cells varies between species. For example, the mosquito has 6 chromosomes, the chimpanzee has 48, and the adders-tongue fern has more than 1,000. We humans have 46 chromosomes within each nucleus in the cells of our bodies. These 46 chromosomes are actually 23 pairs of chromosomes numbered 1 to 23. Chromosome pair 23, the so-called *sex chromosomes*, contains the genes that determine our sex. We have two forms of chromosome 23: X and Y. Females have a pair of X chromosomes (XX), whereas males have one copy of the X and one copy of the Y chromosome (XY). The other 22 pairs of chromosomes are known as *autosomes*.

> **KEY POINT**
>
> Together, the 23 pairs of chromosomes contain our entire instruction manual for life—the genetic code—and each of our cells contains the entire set.

2.3 The instruction manual of life: the genetic code

The *instruction manual of life*, the genetic code, containing all the information that is required to make,

sustain, and proliferate life, is stored in DNA molecules packaged into chromosomes. When our cells

FIGURE 2.5   The packaging of a double-stranded DNA molecule into a chromosome

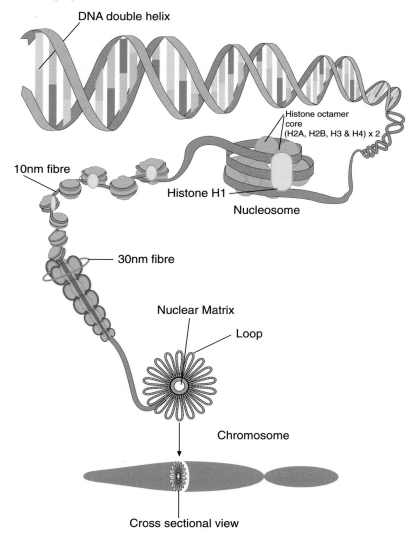

DNA double helix

Histone octamer core (H2A, H2B, H3 & H4) x 2

10nm fibre

Histone H1

Nucleosome

30nm fibre

Nuclear Matrix

Loop

Chromosome

Cross sectional view

require specific proteins the corresponding sections of the genetic code that contain the relevant information, the *genes*, are expressed as described in Box 2.4. Expression of genes means that the coded instructions within the specific genes are copied into RNA molecules (this process is called *transcription*) in the first instance, followed by decoding and protein assembly (this process is known as *translation*).

BOX 2.4

One size fits all!

Everything we are at the chemical and microscopic level is determined by our individual genetic code. This genetic code is held in each of our cells, yet our cells are not all the same. This is because different genes are expressed in different cells. Of course, many genes are expressed or 'switched on' in all cells and these give rise to common proteins that all cells require. The differences between cells arise because they express a different selection of genes and so contain a different array of proteins. A different array of proteins within cells gives rise to different cell structures and different cell functions. However, your cells each still contain all the instructions that would be needed to create a whole new you.

Transcription

For gene expression to occur the information held within the genetic code must move from the nucleus, where it is stored, to the ribosomes in the cytoplasm, where proteins are made. The entire DNA within the nucleus is too big to be effectively moved and it would be too risky to expose the original copy of the genetic code to degradation by certain enzymes present in the cytoplasm. Therefore, the genetic code is first copied and this copy, in the form of an RNA molecule, then carries the information to the ribosomes. This process is called *transcription*.

> ## KEY POINT
>
> Transcription results in the production of complementary RNA molecules that contain copies of sections of the genetic code, which then relocate to the cytoplasm.

Transcribed RNA molecules have exactly the same sequence of bases as the coding strand of the DNA gene sequence, except that thymine (T) is substituted by uracil (U). It is worth noting that not all of the transcribed RNA molecules are destined to be translated into proteins. The cell uses some RNA for other purposes, which are described in Box 2.5. Genes that code for proteins are, however, copied into *messenger RNA (mRNA)* molecules. Enzymes found in the cytoplasm eventually degrade the transcribed RNA molecules. The original genetic code, however, remains intact as DNA within the nucleus, and can be copied into more RNA when next required.

Key features of transcription

Transcription is a complex process that involves a large number of different molecules within the nucleus. We will only outline the key features of the process here. Figure 2.6 shows how all the molecules involved interact during the process of transcription.

- *RNA polymerases* are enzyme complexes that are responsible for producing the complementary RNA transcripts from genes. Humans and all other eukaryotes have three different versions of this enzyme complex: RNA polymerase I, II, and III, each responsible for transcribing different RNA

> ## BOX 2.5
>
> ## Different types of RNA molecules found in cells
>
> There are several different types of RNA molecules in cells, including:
>
> - Ribosomal RNA (rRNA) is used to form part of the ribosome.
>
> - Transfer RNA (tRNA) is used to transfer amino acids to the ribosome, where they are attached to the newly forming protein.
>
> - Messenger RNA (mRNA) contains the copy of the gene code that is used to make proteins.
>
> - Micro RNA (miRNA) is small RNA molecules that bind complementary nucleotide sequences found in target mRNA molecules to form double-stranded RNA. Double-stranded RNA is destroyed by cellular enzymes and therefore miRNAs can regulate the translation of mRNA into protein.
>
> - Small nuclear RNA (snRNA) has a functional role within the nucleus.

molecules. We will specifically look at the role of RNA polymerase II (RNA Pol II), as this enzyme complex, consisting of 12 subunits, is involved in the transcription of the protein-coding genes into mRNA.

- *Promoters* are regulatory sequences of nucleotides within the DNA and act as signals to the nuclear proteins. As they are found upstream of the genes they promote transcription. These signals are diverse in nature and may contain a number of different sections, including:

  - *Initiator (Inr)* sequences, where transcription of a gene starts.

  - *TATA boxes*, which have a sequence of 5′-TATA(A or T)A(A or T)-3′ and are located about 25 to 30 bases upstream from the start point of transcription.

  - *Upstream regulatory elements (UREs)*, which regulate the level of gene expression. Genes that have UREs are more actively expressed than genes without them. These URE sequences are

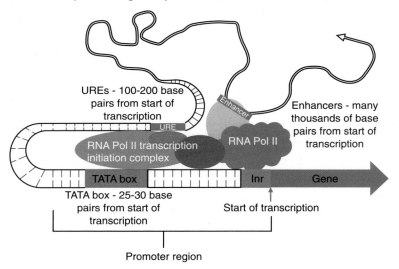

FIGURE 2.6 The RNA Pol II transcription initiation complex and the arrangement of regulatory sequences. RNA Pol II, RNA polymerase II; URE, upstream regulatory elements; Inr, initiator

located 100 to 200 bases upstream of the start point of transcription. Promoters may have one or many UREs and may also have more than one copy of the same URE. *Regulatory proteins* bind to UREs and stimulate the RNA Pol II initiation complex to start transcription.

- *Enhancers* are nucleotide sequences outside of the promoter region. A*ctivator proteins* bind to enhancer sequences and very strongly stimulate expression of their associated genes. These sequences can be about 200 bases in length and can be found many thousands of bases upstream or downstream from the start point of transcription, though they are usually spatially very close to the promoter due to the highly packaged nature of DNA in our nuclei.

- *Transcription factors* are a series of nuclear proteins, including regulatory and activator proteins, which bind to the enhancer and/or promoter sequences of a gene. These transcription factors can then interact with the RNA Pol II enzyme complex to form the RNA Pol II initiation complex to promote the initiation of transcription.

- *Polyadenylation signals* are sequences of nucleotides found at the end of a gene-coding region of DNA. This sequence signals the end of transcription of a gene into mRNA by the RNA Pol II enzyme complex. The corresponding sequence on the mRNA

transcript (5′-AAUAAA-3′) triggers cleavage of the transcript and the addition of a couple of hundred adenine (A) nucleotides (*polyadenylation*) by specific proteins at the 3′ end, forming what is known as the *poly(A) tail*.

Key stages of transcription

Transcription proceeds through three stages:

- *Initiation* involves the formation the *RNA Pol II transcription initiation complex*, consisting of many proteins or transcription factors that associate in a sequential fashion upon the TATA box within the promoter of a gene. Some of these proteins have a role in locally unravelling the packaged DNA, so allowing other proteins to access the promoter. One of the last proteins to join this growing complex is RNA Pol II, which is positioned at the site of the start of transcription. The RNA Pol II transcription initiation complex is shown in Figure 2.6. In highly active genes the regulatory proteins bind to UREs, and activator proteins bind to the enhancer sequences. These regulatory and activator proteins interact directly with and stimulate the RNA Pol II transcription initiation complex to initiate the start of transcription.

- *Elongation* occurs once transcription has been triggered. RNA Pol II separates from the RNA Pol II initiation complex and moves along the template strand of DNA in the 3′ → 5′ direction, unwinding the DNA double helix as it forms the mRNA transcript. Complementary nucleotides are added to the growing mRNA transcript in the 5′ → 3′ direction. The double-stranded DNA re-associates behind the progressing RNA Pol II. The mRNA molecule will have exactly the same base sequence as the coding strand gene sequence of the DNA, except that thymine (T) will be replaced by uracil (U). This mRNA molecule is known as *pre-mRNA* and requires further processing to produce a mature mRNA molecule. This processing includes the addition of a 'chemical cap' (7-methylguanosine) to the 5′ end of the growing pre-mRNA transcript, which protects the molecule from being quickly degraded by other enzymes found in the cell.

- *Termination* occurs when RNA Pol II releases the mRNA transcript. This happens about 1,000 to 2,000 base pairs downstream of the polyadenylation signal (5′-AATAAA-3′) found at the end of a gene. The mechanism of terminating transcription in eukaryotes is not fully understood. The corresponding polyadenylation signal (5′-AAUAAA-3′) that is copied within the mRNA transcript acts as a signal for further modification by polyadenylation, which is the addition of a couple of hundred adenine (A) nucleotides at the 3′ end to form the poly(A) tail. One of the roles of the poly(A) tail is to inhibit rapid degradation of the mRNA by enzymes in the cytoplasm, thereby allowing time for the translation of the mRNA into protein to take place. Some genes contain sequences called *introns*, which do not code for protein. The protein-coding sequences of these genes are known as *exons*. Any intron sequences copied within the mRNA are removed, and the coding sequences are joined up together. This step, known as *splicing*, results in the production of a mature mRNA molecule. The mature mRNA is then transported to the cytoplasm to be translated into protein.

KEY POINT

Transcription is a complex three-phase process whereby genes are copied into RNA molecules.

Translation

Translation is the process that generally follows transcription of mRNA. In translation, the nucleotide sequence, initially transcribed from the genes into an mRNA molecule, is translated into a polypeptide sequence. Before we discuss the process of translation we must first consider how the genetic code is read.

We previously likened the genetic code to a language, and the nucleotides to the letters of that language. The words of the genetic code, however, are only ever written in a sequence of three letters or base sequence triplets called *codons*. There are 64 possible words or codons in the language of our genetic code, as shown in Table 2.1. These codons code either for a single amino acid or for the process of translation to start or stop.

From Table 2.1 you can see that the codon AUG codes for methionine, but it also acts as a 'start' signal for translation. Therefore, methionine is the first amino acid added during translation. Codons UAG, UGA, and UAA, however, all act as 'stop' translation signals and do not code for any amino acid.

The start codon is extremely important because it signals the exact start of the translation process, and so ensures that the codons are read in the correct *reading frame*. Consider a simple mRNA sequence such as 5′-CAUGGCUCUAGUUA-3′. Scanning along the sequence you can see a AUG codon, so using the genetic code, the sequence will be read as 5′-C(AUG)(GCU)(CUA)(GUU)A-3′ and translated as start/methionine-alanine-leucine-valine. If there were not a definitive start signal in this genetic language, the sequence could be read in a number of different reading frames and so interpreted as completely different codons. For example, the codons in our simple sequence could be wrongly read, as 5′-(CAU)(GGC)(UCU)(AGU)UA-3′, translating it into a *mis-sense* polypeptide sequence of histidine-glycine-serine-serine, or it could be read as 5′-CA(UGG)(CUC)(UAG)(UUA)-3′, translating into tryptophan-leucine-stop, producing a *truncated* and incorrect polypeptide.

Key features of translation

Translation occurs when the code within the mRNA molecule is translated into a polypeptide by the

TABLE 2.1 The genetic code

		Second letter				
		U	**C**	**A**	**G**	
First letter	**U**	UUU UUC Phenylalanine / UUA UUG Leucine	UCU UCC UCA UCG Serine	UAU UAC Tyrosine / UAA Stop codon UAG Stop codon	UGU UGC Cysteine / UGA Stop codon UGG Tryptophan	U C A G
	C	CUU CUC CUA CUG Leucine	CCU CCC CCA CCG Proline	CAU CAC Histidine / CAA CAG Glutamine	CGU CGC CGA CGG Arginine	U C A G
	A	AUU AUC AUA Isoleucine / AUG Methionine; Start codon	ACU ACC ACA ACG Threonine	AAU AAC Asparagine / AAA AAG Lysine	AGU AGC Serine / AGA AGG Arginine	U C A G
	G	GUU GUC GUA GUG Valine	GCU GCC GCA GCG Alanine	GAU GAC Aspartic acid / GAA GAG Glutamic acid	GGU GGC GGA GGG Glycine	U C A G

Third letter

ribosomes. Some of the key proteins and molecules involved are outlined here. Figure 2.7 shows the sequence of events in translation.

- *tRNA molecules* have the role of transferring their amino acid cargo to a newly forming protein. There is a specific tRNA for every amino acid and each tRNA molecule has an *amino acid binding site* and an *anticodon sequence*. The anticodon is a triplicate RNA nucleotide sequence that is complementary to the codons in an mRNA molecule. The anticodon therefore specifies which amino acid the tRNA carries. The amino acid binding site on the tRNA is specific for the amino acid coded by the anticodon. The *initiator* tRNA is the first tRNA at the site of translation and always carries methionine. *Ribosomes* consist of a mixture of ribosomal RNA (rRNA) in conjunction with ribosomal proteins. The ribosomes in eukaryotic cells have two parts, a small subunit (known as 40S) and a large subunit (60S), which join together during translation.

- *Initiation factors* are proteins that have a number of roles in initiating translation. For example, they melt the 5′ chemical cap on the mRNA molecule and deliver the mRNA to the ribosomes. Other roles include delivering the initiator tRNA. Initiation

factors are also involved with the formation of a complex between the ribosome subunit, the initiator tRNA, and the mRNA molecule.

KEY POINT

The tRNA molecule with an anticodon sequence complementary to the codon on the mRNA will bring the correct amino acid to the ribosome.

Key stages of translation

Translation, like transcription, proceeds through three phases:

- *Initiation* begins when the initiator tRNA binds to the small 40S ribosome subunit with the help of initiation factors. The 5′ chemical cap on the mRNA molecule is used by other initiation factors to also bind the mRNA molecule to the 40S subunit. The mRNA sequence is then scanned for the first AUG start codon and this triggers the 60S subunit to join the 40S subunit and form the active ribosome. The ribosome starts translation with methionine from the initiator tRNA at the AUG start codon.

- *Elongation* is the process by which the ribosome scans along the mRNA, codon to codon, in

FIGURE 2.7 mRNA translation. (A) Initiation factors modify the mRNA's 5' chemical cap and bind the mRNA to the 40s ribosomal sub-unit. (B) The 40s sub-unit scans the mRNA for the AUG start codon and then joins the 60s ribosomal sub-unit. (C) Initiation: the initiator tRNA delivers methionine (Met). (D) Elongation: the ribosome complex scans along the mRNA in the 5'→ 3' direction. tRNAs bring in the amino acids which link to the growing polypeptide. (E) Termination: the ribosome complex reads a stop codon, the polypeptide is released and the complex dissociates.

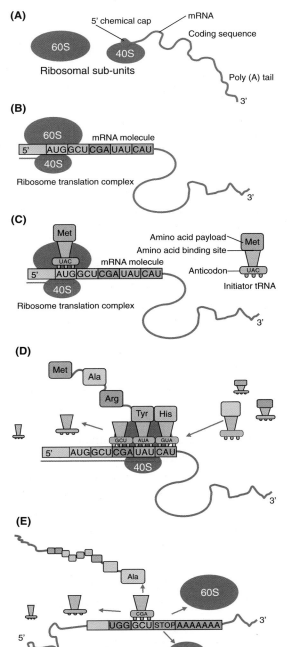

the 5' → 3' direction. Each subsequent codon is matched with the complementary anticodon on the specific tRNA molecules that carry the appropriate amino acid. These amino acids are linked together in a chain by **peptide bonds** to form a polypeptide. After each amino acid is added to the growing polypeptide, the empty tRNA molecule is released from the ribosome.

- *Termination* occurs when the ribosome reaches a stop codon in the mRNA sequence, as no amino acid is brought to the ribosome and the newly synthesized polypeptide is released.

The sequence of events in translation are summarized in Figure 2.7. The synthesized polypeptide chain has to undergo *post-translational modifications* before it becomes a fully functioning protein and is transported to its site of action. Post-translational modifications are alterations to the polypeptide and may include, but are not limited to, hydroxylation, glycosylation, alkylation, and the addition of disulphide bridges or hydrophobic groups. Proteins have diverse functions, from being receptors and enzymes to having structural or signalling roles. Nonetheless, they are all produced in a similar way.

 Have a look at Section 3.1 'Structure and function of proteins', in Chapter 3, for more information on the topic.

KEY POINT

The sequence of the amino acids in a protein polypeptide is dictated by the codon sequence of the mRNA, which, of course, is dictated by our genes.

SELF CHECK 2.4

Can you determine the polypeptide sequence that is coded for by the mature mRNA sequence below?

5'-GUAUCCGAGAGAGGACAUGGAUCAGGGUUGUGA
AGGAGAUUACAAAAGAUGGUGUGAUAUAGAGUAUCUG
GGAGAAAACUAGAGGAAUAAAAAAAAAAAAAAA
AAAAAAAAAAAAAAAAA-3'

31

2.4 How is our genetic code protected and passed on?

So far we have looked at our genetic code in terms of how it is used by our cells to produce different RNA molecules and, of course, proteins. However, the genetic code also has to be passed on from parent to daughter cells during cell division. This process starts at the very beginning of our early development in the womb and continues until our death. As we grow, our cells divide and differentiate to produce the different tissues, organs, and body systems. Even as adults, many of our cells still have to divide when a tissue needs to be repaired or when old cells need to be replaced. Our genetic code is passed down through many generations of cells. Before we consider how a cell facilitates this transfer of genetic information we must first understand the process of cell division.

> **KEY POINT**
>
> It is imperative that our cells are able to accurately duplicate their genetic code to pass on to their daughter cells, so each cell contains an identical copy of the full genetic code.

The cell cycle

The cell cycle could, perhaps, be better described as the cell 'division' cycle, as this is the sequence of events that occurs when a cell is triggered to divide. It is termed a 'cycle' because these events are repeated every time cells divide.

The series of events in the cell cycle are shown in Figure 2.8 and fall into four broad phases:

1. *Gap 1 (G_1)*: The cell prepares itself to make a new copy of its DNA. Genes that encode proteins and enzymes involved in the synthesis of new DNA are switched on.

2. DNA *Synthesis (S)*: The cell synthesizes a copy of its entire genetic code, resulting in duplication of every pair of chromosomes. This process is called *DNA replication* and we will look at this under the next heading of the chapter.

3. *Gap 2 (G_2)*: The cell prepares for cell division. Genes coding for microtubules are switched on. Microtubules are needed during the act of cell division and are also required by the daughter cells for cell structure.

4. *Mitosis (M)*: The cell divides into two duplicate daughter cells. These daughter cells, when the time is right, will begin the cycle again from the G_1 phase.

Some cells temporarily or permanently drop out of the cell cycle during the G_1 phase. These cells are called *quiescent* cells and are in what is known as the G_0 phase. Cells can drop out of the cell cycle to preserve energy or if the cell has sustained damage to its DNA.

> **KEY POINT**
>
> Most of our cells divide numerous times during our life; however, some cell types, such as nerve and heart cells, once mature do not divide, and are with us for life.

FIGURE 2.8 **The cell cycle**

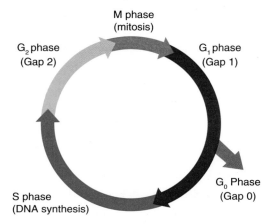

M phase
(mitosis)

G_2 phase
(Gap 2)

G_1 phase
(Gap 1)

G_0 Phase
(Gap 0)

S phase
(DNA synthesis)

DNA replication

The process of DNA replication takes place during the S phase of the cell cycle and is a highly regulated event. Each of our chromosomes has a double helix DNA molecule that may be hundreds of millions of base pairs in length, which needs to be replicated with minimal error. The general process of eukaryotic replication is illustrated in Figure 2.9.

Replication of each chromosomal DNA molecule begins at sites within the DNA sequence known as *origins of replication*, and proceeds in both directions away from these origins. In eukaryotes, a single chromosome may contain several thousand origins of replication. To initiate replication, a protein complex called the *origin recognition complex (ORC)* must bind to each of the origin of replication sites. The ORC then recruits a series of other proteins, including *helicases* and *topoisomerases*. Helicases are enzymes that function to unwind the double helix, locally opening up the DNA to allow replication to occur. However, this results in increased coiling ahead of the replication fork. Topoisomerases therefore work ahead of the replication fork by cutting, uncoiling, and resealing the DNA strand. The complex consisting of the ORC, helicases, and some other proteins is known as the *pre-replication complex*. This complex needs to be triggered by a protein known as the *licensing factor* in order for replication to be initiated. The licensing factor ensures that replication occurs at precisely the right time, and only once during the cell cycle.

When the helicases in the pre-replication complex locally unwind the double helix, both strands of DNA become accessible and they both act as templates. The point where the strands separate is known as the *replication fork*. This leaves the nucleotide bases of each strand temporarily exposed so that they can base pair with 'new' complementary nucleotide bases as replication proceeds. A family of enzymes called *DNA polymerases (Pol)* carry out this process. In eukaryotes this family consists of Pol α, Pol ε, and Pol δ, each with distinct roles. At the replication fork, due to the antiparallel nature of DNA, one of the template strands runs in the $5' \rightarrow 3'$ direction, while the other runs in the $3' \rightarrow 5'$ direction. However, DNA polymerase is only capable of synthesizing a new strand of DNA in the $5' \rightarrow 3'$ direction, and therein lies a problem. The *leading template strand* runs in the $3' \rightarrow 5'$ direction and so its new complementary strand has a $5' \rightarrow 3'$ direction and can be synthesized in one continuous run, following the replication fork as it progresses and opens up the DNA. However, the *lagging template strand* runs in the $5' \rightarrow 3'$ direction, so its new complementary strand has to be synthesized in the opposite direction to the progression of the replication fork. To circumvent this problem, the DNA polymerase that is copying the lagging strand synthesizes small DNA fragments known as *Okazaki fragments*.

Replication starts by the addition of *RNA primers*—short strands of RNA that are complementary to a short DNA sequence in the template strand. These primers are made and added to the template DNA strands by Pol α. The leading strand requires only one primer, but Pol α continually adds primers to the lagging strand as the replication fork opens up. Pol ε uses the RNA primer as an anchor to start adding complementary nucleotides to the leading strand in a continuous run. Pol δ fills in the gaps between the primers with complementary nucleotides and replaces the primer RNA with DNA. Adjacent Okazaki fragments are finally joined with phosphodiester bonds by the enzyme *DNA ligase*, creating the new DNA strand that is complementary to the lagging template strand.

KEY POINT

DNA polymerase can only synthesize a new DNA strand in the $5' \rightarrow 3'$ direction.

There may be several thousand active replication forks within a single chromosome, all simultaneously copying the whole DNA molecule. When all the replication forks meet, the result is two identical complete copies of the chromosome's linear, double-stranded DNA molecule.

SELF CHECK 2.5

Why are Okazaki fragments formed in during eukaryotic cell DNA replication?

FIGURE 2.9 Eukaryotic replication

(A)

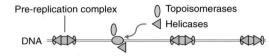

Origin recognition complexes (ORC) bind to origin of replication sites found within the DNA molecule

(B)

The ORC recruits other proteins including topoisomerases and helicases to form the pre-replication complex

(C)

Replication forks

The helicases then open up the DNA forming replication forks to allow access for DNA polymerases. Topoisomerases cut and re-seal the DNA to relieve the tension formed in the DNA

(D)

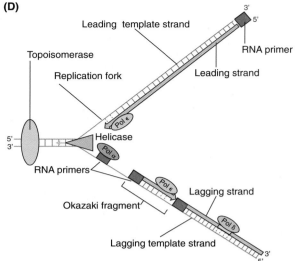

The leading template strand is copied in one continuous daughter leading strand. The lagging strand is synthesized in a discontinuous fashion via the production of Okazaki fragments. The primers are replaced with the appropriate DNA nucleotides and ligases seal any gaps with phosphodiester bonds

2.5 What can happen when things go wrong?

During replication, DNA polymerase occasionally adds or removes nucleotides from a sequence, or adds the wrong nucleotide. This is thought to occur at a rate of about 1 in 100,000 nucleotides, which may seem pretty rare, but results in more than 100,000 mistakes being made every time a cell divides. This level of error would be unsustainable, as it would result in too significant a number of changes in the genetic code. Our cells have therefore developed mechanisms to repair damaged DNA, and most, but not all, of these mistakes are mended. During replication, DNA polymerases, in addition to duplicating the DNA molecule, carry out *proofreading* functions and thus correct the majority of faults. In addition, after replication, *mismatch repair enzymes* examine the newly formed double-stranded DNA for mismatched nucleotide base pairs and replace the wrong nucleotide with the correct one.

In addition to errors caused during replication, the cells of our bodies are also constantly exposed to environmental factors that damage our DNA, some of which we will consider in 'Causes of mutations' within this section. If the damage or mistakes in our DNA are not repaired, then changes in the DNA sequence of our genetic code become permanent and could be passed on to daughter cells. These permanent changes are known as *mutations*. If the mutations occur in **germ cells** they can be passed on from ancestors through generations and generations, leading to *inherited genetic mutations*. Some of these inherited genetic mutations result in diseases, others may be harmless, and some may even be advantageous, as is explained in Box 2.6.

Causes of mutations

The environmental causes of DNA damage that can lead to mutations mostly fall into one of three broad groups:

- *Physical causes* include ultraviolet (UV) light, which is part of the sunshine we all enjoy, and ionizing

Not all mutations are bad

Some good or advantageous mutations can exist in our genetic code and are passed on for generations. Sickle cell anaemia (SCA), an inherited disease affecting the red blood cells, is an example of when certain mutations can be advantageous.

Malaria can be lethal and, before modern medicine and procedures for mosquito control, it was widespread in West Africa. When the British Empire first colonized West Africa, thousands of the British died from malaria. West Africa was nicknamed 'the white man's grave'. Yet the West Africans themselves did not die in such numbers. This was because SCA was prevalent in West Africans. Those that had two copies of the chromosome with the sickle cell mutation had the disease and were often ill and died young. Those, however, that had one normal and one mutated chromosome were *carriers* of SCA. In genetic terms, carriers are those people who carry a copy of the mutated gene but do not show or have only mild symptoms of the disease. Therefore, these SCA carriers did not have the disorder themselves. However, these individuals were found to be resistant to malaria. These carriers were more likely to survive to adulthood and pass on the advantageous SCA gene to their children. Those that inherited normal chromosomes were susceptible to malaria and were less likely to survive to adulthood and have children. This is how an advantageous mutation can be passed on from our ancestors and can persist in our genetic code.

radiation. UV light has a higher energy than visible light, and can penetrate our skin cells and damage DNA. UV light can cause new chemical bonds or cross-links to form between adjacent cytosine and thymine bases, forming *pyrimidine dimers,* or create free radicals that can damage DNA. It is the damage to our DNA from UV light that gives rise to sunburn. If this DNA damage is too severe to be fixed, our cells self-destruct in a process called *apoptosis*. This results in our skin peeling off and being replaced by new skin. Apoptosis therefore can help to protect us against mutations. Ionizing radiation such as X-rays, gamma rays, and cosmic radiation from space have enough energy to penetrate our entire bodies and cause damage to the DNA in all of our cells. This damage can include single- or double-strand breaks in the chromosomal DNA molecule, and destruction of bases or sugars in nucleotides.

- *Chemical causes* can come from the environment we live in, such as from cigarette smoke or pollution. People who work in certain industries may be exposed to DNA-damaging chemicals, such as the arylamines found in some dyes used in textile production, and the polycyclic aromatic hydrocarbons used in the petroleum industry. The food we eat may also contain DNA-damaging agents: well-cooked meat and sodium nitrate, used as a food preservative, have been associated with DNA damage. Many of these agents that we are exposed to do not directly damage DNA but are broken down by the liver to produce DNA-damaging byproducts. For example, spices, cereals, beans, nuts, and other crops, if stored incorrectly, may become contaminated with aflatoxin, produced by the fungus *Aspergillus*. Aflatoxin is metabolized by the liver to produce a powerful mutagen and carcinogen. Chemical damage to DNA includes single- or double-strand breaks, bond formation between adjacent bases or between DNA strands, the addition of chemical groups to bases, and the removal of bases.

- *Biological causes* primarily involve cellular parasites such as **viruses**. The majority of viruses cause little or no permanent harm to us, but a few are associated with serious diseases and **cancers**. During infection, viruses take command of the host's replication machinery to replicate their own genes. Their DNA is inserted into our own, often causing mutations to our genes, which may destroy the genes or alter the activity of the gene products. DNA mutated by the insertion of viral DNA might not directly result in problems for the cell, but the proteins expressed from the viral DNA can. Viruses also commandeer other cellular proteins to assemble or package more viruses in order to infect other cells.

DNA damage can also occur without the presence of environmental causative factors. It can be caused in error by normal cellular enzymes and proteins during cell division and, if left unrepaired, can lead to spontaneous mutations.

Our cells can respond to and repair DNA damage, thus minimizing the effect it may cause. In addition to the mismatch repair enzymes mentioned before, during any part of the cell cycle, *excision repair enzymes* inspect our DNA for damage. When damage is discovered, these enzymes remove the damaged DNA and DNA polymerase replaces the missing nucleotides.

Probably the most important protein that is activated by DNA damage is called *tumour protein 53* or just *P53*. P53 is known as the 'guardian of the genome' for its role in protecting the integrity of our genetic code. P53 halts the cell cycle in the G_1 phase just before the DNA synthesis (S) phase (at the G_1/S checkpoint). If DNA damage is detected, P53 activates the excision repair enzymes to repair the damage. If the damage is successfully repaired, the cell is allowed to continue through the cell cycle; if not, P53 triggers the cell to destroy itself by undergoing apoptosis. In this way, P53 prevents mutations being passed on to daughter cells. However, despite the mechanisms our cells have to protect the fidelity of our genetic code, mutations are inevitable.

Types of mutations

Unrepaired DNA damage can give rise to a number of different types of mutations as detailed below:

- *Substitution* or *point* mutations are caused if a single nucleotide base is changed in the DNA sequence. There are two types of substitution mutation: *transition* and *transversion* mutations. In a transition mutation, a purine base is changed for another purine base (A to G or G to A) or a pyrimidine base is altered to another pyrimidine base (C to T or T to C). In a transversion mutation, a pyrimidine is substituted by a purine or vice versa.

- *Frameshift* mutations are caused when the reading frame of the gene is altered. This is the consequence of nucleotides being added to or removed from the DNA. Mutations that occur when a nucleotide or a number of nucleotides are added to a DNA

sequence are called *insertion* mutations, and mutations where they are removed from the sequence are called *deletion* mutations.

- *Chromosomal translocations* occur when a fragment from one chromosome is incorrectly added to another chromosome.

Consequences of mutations

So far we have looked at how mutations might arise in cells, but what are the consequences of getting mutations in the genetic code? The consequences are largely dependent on which and how many genes are affected, in addition to the location of the mutation within the gene. In the case of a substitution mutation there might not be any consequences. For example, if the codon GC*U* were mutated to GC*C*, there would be no change in the amino acid sequence of the protein as they both code for the same amino acid (alanine). On the other hand, substitution mutations can result in considerable consequences if the change produces a stop codon in the middle of a protein-coding sequence, or codes for an incorrect amino acid to be inserted into a significant part of the polypeptide chain. For example, a substitution mutation of the 6th codon in the sequence of the β-globin gene results in glutamic acid being changed to valine. This change in the amino acid in the β-globin protein gives rise to *sickle cell anaemia*. Another substitution mutation within the β-globin gene at the 39th codon forms a stop codon (*CAG* is mutated to *UAG*), resulting in a useless short protein chain that does not function as a β-globin protein. This gives rise to the clinical symptoms of β-thalassaemia.

Single gene disorders

Inherited genetic diseases affecting single genes fall into two categories: *recessive* and *dominant*. Recessive diseases require the mutation to be present in both copies of a chromosome to generate the most severe disease state, whereas dominant mutations need the mutation in only one chromosome for there to be clinical symptoms. Based on the chromosome type that bears the mutation, there may be autosomal or sex chromosome-linked (both X and Y) disorders. Many genetic diseases exist that result in a

TABLE 2.2 Genetic diseases that result from inherited mutations

Inheritance trait	Disease	Mutation type	
Autosomal recessive	Cystic fibrosis	Frameshift (deletion)	
	Sickle cell anaemia	Substitution (transversion)	
	Tay-Sachs disease	Numerous types of mutation give rise to disease	
	β-Thalassaemia	Substitution (transition)	
Autosomal dominant	Lynch syndrome (hereditary non-polyposis colorectal cancer)	Can result from mutation in a number of different DNA mismatch repair enzyme genes	
	Huntington's disease	Frameshift (insertion of multiple nucleotides)	
	Hypercholesterolaemia, type B	Numerous types of mutation give rise to disease	
X-linked recessive	Duchenne muscular dystrophy	Frameshift (deletion)	
	Haemophilia	Numerous types of mutation give rise to disease	
X-linked dominant	Hypophosphataemic rickets	Frameshift (deletion)	

diverse range of clinical symptoms. Some examples of genetic diseases resulting from mutations are given in Table 2.2.

An example of an autosomal recessive single gene disorder is cystic fibrosis (CF). CF is particularly prevalent in Caucasians. A mutation within the cystic fibrosis transmembrane regulator (CFTR) gene in both copies of chromosome 7 affects the secreting glands in the body, particularly in the lungs. If a child inherits a mutated copy of the CFTR gene from one parent and a normal copy from the other parent they will be symptom free; however, they will also be a CF *carrier*, meaning they may pass on the genetic disorder to their offspring. About 1 in 25 people in the UK are estimated to be CF carriers. Children born to parents who are both CF carriers might inherit the mutated chromosome 7 from each parent and have CF (there is a 25% chance of this occurring). They could, however, inherit the normal chromosome 7 from both parents and neither have CF nor carry the defective gene to pass on to the next generation (also a 25% chance). Case Study 2.1 further explores the inheritance pattern of CF. Sickle cell anaemia, described earlier in Box 2.6, is also a recessive single gene disease.

Some inherited single gene disorders are sex chromosome-linked. For example, a mutation in the dystrophin gene found on the X chromosome results in the muscle-wasting disease Duchenne muscular dystrophy.

The disease is carried by females and is considered to be recessive. However, as males have only one copy of the X chromosome the disease is prevalent in males; females have two X chromosomes and therefore can have a normal copy of the dystrophin gene to compensate for the faulty one. If a carrier has a child, the child has 50:50 chance of receiving the mutated X chromosome from the mother. Therefore, if the child is female she will be a carrier, if male he will have the disease.

Although many genetic diseases are caused by inherited mutations, some are also caused by spontaneous mutations. Duchenne muscular dystrophy may also occur due to spontaneous mutations in the dystrophin gene. Often, no familial genetic link is evident and the disease is considered to have arisen by spontaneous mutation during prenatal development. Of course, this means that it is also possible that, during prenatal development, spontaneous mutations could occur in the normal copy of a gene in recessive disorders such as CF, giving rise to disease.

SELF CHECK 2.6

Although X chromosome-linked recessive genetic diseases such as Duchenne muscular dystrophy are predominant in males, they are not exclusively male diseases. Sometimes, females also have symptoms of the disease. How could this be?

Other genetic disorders

Some diseases can come about from inherited mutations in more than one gene. These are considered to be *multiple gene disorders*. Lynch syndrome is an autosomal dominant genetic disease caused by mutations in a number of the mismatch repair enzymes' genes. Approximately 4 out of 5 people with Lynch syndrome develop cancer of the digestive tract during their lifetime and so is often called hereditary non-polyposis colorectal cancer.

Humans normally have 46 (23 pairs) chromosomes; however, genetic diseases can come about when an incorrect number of chromosomes are erroneously packaged into an egg or sperm and passed on to a child. These are called *chromosomal disorders* and can result in cells having a missing chromosome

(*monosomy*) or an extra chromosome (*trisomy*). For example, Down's syndrome results from an extra copy of chromosome 21. Trisomy can also happen with the sex chromosomes. A female-only condition called triple X syndrome is caused by an extra copy of the X chromosome (XXX). This disease usually results in fairly mild symptoms. Klinefelter's syndrome is a disease that occurs in males who have an extra X chromosome (XXY). Although not life-threatening, the disease can cause infertility and other problems.

Nowadays, people can have genetic tests to find out if they are carriers of genetic diseases such as CF or SCA. Case Study 2.1 illustrates how, as a result of the availability of genetic testing, a couple can find out whether their unborn child has a genetic disease or not.

CASE STUDY 2.1

Angela has told Ravi she is pregnant and they are both thrilled at the prospect of being parents. However, Angela had an older brother who had cystic fibrosis (CF) and died aged 25 and she does not want this for her child. Angela knows that she is a carrier of CF but Ravi cannot remember if there is a history of CF in his family. They are both worried there may be a chance their baby inherits CF.

REFLECTION QUESTIONS

1. How can Ravi and Angela find out if Ravi is a carrier for cystic fibrosis?

2. If Angela is a carrier of cystic fibrosis and if Ravi's test proves negative, why isn't her baby going to have cystic fibrosis?

3. If Ravi also carried the same cystic fibrosis gene mutation as Angela, would the baby have cystic fibrosis?

Answers

1 Ravi will need to contact his GP and arrange to have a cystic fibrosis carrier test. In this simple test, Ravi will be asked to use a mouthwash to rinse his mouth out and the mouthwash will be collected and sent to the laboratory for analysis. The mouthwash will contain enough of Ravi's cells for the laboratory to extract his DNA to test for the presence of the cystic fibrosis mutation.

2 Our chromosomes come in pairs. That means that we have two copies of each. Cystic fibrosis is a recessive disorder. As a carrier, Angela will have one copy of chromosome 7 with a cystic fibrosis gene mutation, and one copy of chromosome 7 with the normal gene. The baby receives one chromosome from each parent. Therefore there is a 50% chance the baby will receive either the chromosome with the normal or the mutated gene from Angela. If Ravi is not a carrier for the genetic disorder the baby cannot inherit cystic fibrosis as it can only inherit a normal copy of the gene from Ravi. The baby can however be a carrier for the disorder, the same as Angela.

3 If they are both cystic fibrosis gene mutation carriers, Ravi and Angela would each have a normal and a mutated version of the gene on chromosome 7. If the baby had received a chromosome with the mutated gene from both Ravi and Angela, the baby would have cystic fibrosis (1 in 4 chance, 25%). If only one of them had passed on a chromosome with the mutated gene the baby would be a carrier of cystic fibrosis (1 in 2 chance, 50%). If neither of them passed on the chromosome with the defective gene, the baby would not have nor carry cystic fibrosis (also 1 in 4 chance, 25%).

Cancer

Cancer is a disease that can arise from inherited mutations. Many breast, ovarian, and colon cancers are known to be associated with inherited mutations. However, many diseases result from sporadic mutations as a result of DNA damage. The majority of cancers are caused by these sporadic mutations. For example, most cases of chronic myelogenous leukaemia and some other leukaemias are caused by a chromosomal mutation—the *Philadelphia chromosome* (so called because it was discovered in Philadelphia, Pennsylvania, USA). The Philadelphia chromosome results from a fragment of chromosome 9 breaking off and joining onto chromosome 22 and vice versa (a process known as *translocation*). Chronic myelogenous leukaemia is characterized by unregulated cell proliferation in the bone marrow.

There are only a handful of viruses that are thought to cause cancer and, collectively, are believed to give rise to about 1 in every 6 cancers. For example, Epstein–Barr virus has been linked to cancers such as Hodgkin's lymphoma and Burkitt's lymphoma. Infection with either hepatitis B or C virus in some people is thought to trigger liver cancer. Some strains of the human papilloma virus (HPV) cause harmless common warts; however, other strains cause cervical cancer and virtually every case of cervical cancer is a direct result of HPV infection. This virus is sexually transmitted and infects the cervical cells. The expressed viral proteins hijack P53, the guardian of the genome (see 'Causes of mutations' within this section), and other cell cycle control proteins. As a result of this, the virus is able to drive uncontrolled cell division. In the UK, schoolgirls aged 12 to 13 years are offered a vaccine against HPV that is effective at preventing HPV infection.

> **SELF CHECK 2.7**
>
> Why are schoolgirls offered the HPV vaccine at 12 to 13 years of age?

Current and future genetic therapies

We have seen how disease can arise because of changes or mutations made to our genetic code.

Current treatments for some of these disorders only address the symptoms and not the underlying cause. Scientists have now, through the work of the Human Genome Project, decoded the entire 3 billion or so base pairs that make up our genetic code. So is it possible that scientists can use this knowledge to create medicines to cure genetic diseases? The answer is maybe, and research to this end is currently ongoing.

One approach currently being explored is the silencing of genes that are over-expressed in diseases. To silence or suppress an overactive gene, the expressed mRNA is targeted for destruction and so less mRNA is translated into protein. Short interfering RNA (siRNA) molecules are designed to bind to the complementary nucleotide sequence found in the target mRNA and, in a similar way to that of endogenous miRNA molecules, label the mRNA for destruction by cellular enzymes. This general approach is known as RNA interference (RNAi) and research using RNAi is now being carried out in the search for possible treatments for diseases such as cancer.

'Fixing' a mutation is the central goal of *gene therapy*. Gene therapy involves administering *therapeutic DNA* which, when expressed by the cell, acts as a substitute for the mutations present in the patient. It was used successfully for the first time in 1990 to treat a 4-year-old girl who had severe combined immunodeficiency disease (SCID). People with this very rare disease have a very poor or non-existent immune system owing to a mutation in the gene for an enzyme called adenosine deaminase (ADA). Scientists inserted a normal copy of the ADA gene into the girl's white blood cells. The 'fixed' cells were propagated to increase their numbers and then administered to the child. When these fixed cells then divided, their daughter cells carried the new gene. The girl regained an immune response and the gene therapy was considered a success. Since then, a number of patients with SCID have been successfully treated with gene therapy.

This initial success in gene therapy triggered researchers to search for other applications and methods in the treatment of genetic diseases. The main problem, however, is how to get a correct copy of a gene into the DNA of enough cells to relieve the symptoms or cure the disease. A number of approaches to this problem have been investigated. One such idea was to add a therapeutic gene into a virus. Logically, viruses should make excellent carriers for therapeutic DNA because that is what they do—invade cells and add their DNA into the host cell's DNA. This approach is extremely efficient in the laboratory, but has proved problematic in humans because our immune system can mount an immune response to the viral carrier. In the USA in 1999, a healthy teenage boy died taking part in a clinical trial for a gene therapy treatment when he had a severe reaction to the viral therapeutic gene carrier used in the trial. Subsequently, it was found that the research practices of those involved in the trial were questionable and unethical, and this has influenced the revision of the guidelines and documentation involving clinical trials. This outcome was of course very sad for the boy's family, but was also a massive setback for gene therapy research.

 The ethics of treatment will be discussed in Chapter 3 'Legal and ethical matters', in the **Pharmacy Practice** book of this series.

Scientists are also trying other, non-viral, means of delivering therapeutic DNA that are less likely to illicit immune responses in recipients. These include the use of liposomes and polymers as vehicles. Future research will undoubtedly result in permanent cures for some of the genetic disorders that conventional medicines today can only treat symptomatically.

KEY POINT

Gene therapy has the potential to actually cure a genetic disease whereas current therapies can only treat the symptoms.

CHAPTER SUMMARY

➤ Cells can be divided into prokaryotes and eukaryotes.

➤ Eukaryotic multicellular organisms like humans exist because cells differentiate into specialist cells that make up the tissues and organs that contribute to the whole organism.

➤ The basic units of DNA and RNA molecules are the nucleotides.

➤ The 'letters' of the genetic code come from the bases guanine (G), cytosine (C), adenine (A), and thymine (T) within DNA. In RNA, uracil (U) replaces thymine (T).

➤ The 'words' of the genetic code are written in base sequence triplets called codons.

➤ DNA has to be highly packaged into chromosomes to fit into the nucleus.

➤ Coding genes have base sequences that promote transcription.

➤ Enzymes called DNA polymerases carry out transcription of DNA to make RNA.

➤ Translation of mRNA into protein occurs at the ribosomes.

➤ The cell cycle has four phases, known as the G_1 (gap 1), S (synthesis), G_2 (gap 2), and M (mitosis) phases.

➤ DNA replication occurs during the S phase.

➤ DNA damage can occur because of mistakes during DNA replication, or from physical, chemical, or biological damage.

➤ Mutations occur when unrepaired DNA damage is passed on to daughter cells, and can be advantageous, harmless, or cause disease.

➤ Mutations in germ cells can be passed down through generations.

➤ Chromosomal problems can result in disease.

➤ The treatment of some inherited disorders may lie within gene therapy.

Elliott, W.H. and Elliott, D.C. *Biochemistry and Molecular Biology*. 4th edn. Oxford University Press, 2009.

This is a well-written textbook that covers molecular biology in depth.

King, R.J.B. and Robins, M.W. *Cancer Biology*. 3rd edn. Pearson Prentice Hall, 2006.

This textbook will give you a detailed understanding of how mutations can lead to cancer.

The biochemistry of cells

HELEN BURRELL

As a pharmacist you will be responsible for the supply of drugs. It may seem common sense that, to do this safely, you must know about the chemical composition of the drug in order to make a judgement about the effects it could have on the human body. However, it is equally important to know and understand the basic human chemical processes, as drugs can affect them. These processes explain why some unwanted side effects arise when taking certain medication, and they need to be fully understood to improve drug delivery. The study of these chemical processes in living organisms is called *biochemistry*. Biochemistry deals with the structure and function of all the main cellular components, including proteins, carbohydrates, lipids, and nucleic acids.

In biological systems, proteins are involved in a diverse range of vital cellular processes; this chapter discusses the function of proteins with respect to their structure and levels of organization. It then focuses on one major type of proteins, the enzymes, and their regulation. Enzymes are of particular importance to the pharmaceutical industry as many drugs are enzyme inhibitors. This chapter therefore discusses the kinetics of enzyme activity and the pathways by which they can be inhibited or activated. One of the key functions of enzymes is to regulate metabolism. Think about what happens when you eat some food after feeling tired and hungry. A little while after, you start to feel less tired and more energetic. This happens more quickly if you eat something sugary. This is all because of changes in metabolism. The stages of metabolism are outlined in this chapter, with particular reference to the enzymes that catalyse each step and how the key stages are regulated. As glucose is the preferred fuel for the brain and red blood cells, this chapter raises the question of what happens when the glucose level is low, and considers the alternative fuels used by our bodies.

Learning objectives

After reading this chapter you should be able to:

- Define the three types of protein and give examples of where they can be found in the body.

- Differentiate between the primary, secondary, tertiary, and quaternary structure of a protein and identify the forces or bonds that help to stabilize each structure.

- Describe what an active site is within an enzyme and how it is structurally formed.

- Give examples of the four types of reversible enzyme inhibitor and describe how they affect K_m and V_{max}.

- Describe how the three types of irreversible inhibitor affect enzymes.

- Understand what happens to the three major fuels (carbohydrates, fatty acids, and proteins) within the body in terms of where they enter the metabolic pathways and how they are regulated.

3.1 Structure and function of proteins

Proteins can generally be classified into three groups according to their shape and solubility:

1. *Globular proteins* are almost spherical and are highly soluble in aqueous solutions. Their solubility is due to the positions of hydrophobic and hydrophilic amino acids within the protein. The hydrophilic amino acids are on the outside, where they interact with the aqueous environment, whereas the hydrophobic amino acids are protected on the inside of the molecule. Globular proteins have many roles: they include enzymes with catalytic activity, for example acetylcholinesterase; transportation devices to aid the passage of molecules through membranes, for example potassium channels; messenger signals such as hormones, for example insulin; or they can be regulatory molecules, for example histones or transcription factors. These proteins are easily denatured.

2. *Fibrous proteins* are linear proteins that are insoluble in aqueous solutions. They form aggregates owing to the hydrophobic side chains on the outside of the molecule. They are involved in cell structure to provide protection and support, and as a consequence they are components of connective tissue, tendons, muscle fibres, and the bone matrix. Examples of fibrous proteins include keratin, collagen, and elastin. In general, fibrous proteins are not denatured as easily as globular proteins.

3. *Membrane proteins* are integral, peripheral, or lipid-anchored proteins associated with cell membranes, and each has different functions. Many are receptors for endogenous and therapeutic molecules. They have hydrophobic amino acids pointing outwards from the protein molecule, which enable interaction with the non-polar phase of the membrane, and have fewer hydrophilic amino acids than globular proteins. They are generally insoluble in aqueous solutions unless a detergent is present.

Although there are different classifications of proteins, all protein structures have similarities. There are four levels of protein structure: the *primary*, *secondary*, *tertiary*, and *quaternary* levels.

SELF CHECK 3.1

What are the differences between the main types of protein?

Primary structure

Proteins are made up of units called amino acids and the sequence of these in any given protein is called the *primary structure*.

Although amino acids (except for glycine) can form two enantiomers (D and L), only L-2-amino acids (also called L-α-amino acids) are found within proteins. Proteins are predominantly made up of the 20 common amino acids shown in Table 3.1, but there are a few uncommon amino acids that are found in specific proteins (for example hydroxylysine and hydroxyproline). Each protein has its own unique amino acid sequence, which is specified by individual genes and is created during translation.

 You can read more about translation in Section 2.3 'The instruction manual of life: the genetic code', within Chapter 2 of this book.

The primary structure of a protein is the level at which the amino acids are joined together by *peptide bonds* (also called amide bonds). This forms a linear polymer known as a polypeptide. Each amino acid within the polypeptide is called a *residue*, and proteins range in size from tens to thousands of residues.

 Chapter 9 'Chemistry of biologically important molecules', in the **Pharmaceutical Chemistry** book within this series, has more in-depth coverage of amino acid chemistry and peptide bond formation.

Each end of a polypeptide chain is different: there is an *amino group* at one end and a *carboxyl group* at the other. Proteins are therefore said to have *polarity*. Scientists use the convention of the amino group defining the start of the polypeptide chain and so the amino acid with the free amino group is written first. A polypeptide chain consists of a regularly repeating part called the *backbone*, and variable parts called *side chains*. Each amino acid residue has a carbonyl group (C = O) and, with the exception of proline, an amine (NH) group. Whereas the carbonyl group is a good **hydrogen bond acceptor**,

TABLE 3.1 Nomenclature and properties of the 20 common amino acids

Amino acid	Three-letter abbreviation	One-letter abbreviation	Properties
Alanine	Ala	A	Neutral non-polar
Arginine	Arg	R	Basic polar
Aspartate	Asp	D	Acidic polar
Asparagine	Asn	N	Neutral polar
Cysteine	Cys	C	Neutral slightly polar
Glutamate	Glu	E	Acidic polar
Glutamine	Gln	Q	Neutral polar
Glycine	Gly	G	Neutral non-polar
Histidine	His	H	Basic polar
Isoleucine	Ile	I	Neutral non-polar
Leucine	Leu	L	Neutral non-polar
Lysine	Lys	K	Basic polar
Methionine	Met	M	Neutral non-polar
Phenylalanine	Phe	F	Neutral non-polar
Proline	Pro	P	Neutral non-polar
Serine	Ser	S	Neutral polar
Threonine	Thr	T	Neutral polar
Tryptophan	Trp	W	Neutral slightly polar
Tyrosine	Tyr	Y	Neutral polar
Valine	Val	V	Neutral non-polar

the amine group is a good **hydrogen bond donor**. It is these groups that interact with each other or the functional groups from side chains in order to stabilize the protein structure.

In some proteins the linear polypeptide chain is cross-linked. The most common cross-links are called *disulphide bonds*. These are formed when a pair of **cysteine** residues is oxidized to form a cystine unit, as shown in Figure 3.1(A). These disulphide bonds are commonly found in extracellular proteins and are usually lacking from intracellular proteins. *Non-disulphide bonds* are a further, rare, form of cross-link. These form between side chains other than cysteines, for example lysines, as shown in Figure 3.1(B). Here, the lysine derivative allysine is produced by the action of the enzyme lysyl oxidase; a pair of allysine residues can then form an aldol cross-link. These can be found in specific proteins such as collagen in connective tissues and fibrin in blood clots and are present to increase strength in the protein structure.

KEY POINT

The primary structure of proteins is their specific sequence of amino acids.

We need to know the primary structure of proteins to:

- Identify when alterations have occurred. Mutations in the genetic code can result in a single amino acid being changed and this in turn can cause severe diseases including cystic fibrosis or sickle cell anaemia.

- Produce a three-dimensional model of the protein. Folding enables the protein to adopt a *functional shape*. Folding of proteins is highly critical as misfolding can cause diseases such as Parkinson's or Alzheimer's.

- Identify protein function from the primary structure. For example, if the protein is an enzyme the primary structure will help determine the catalytic mechanism.

- Identify the evolution of a protein, because similarities are the result of common ancestry.

 You can read more about genetic disorders in Section 2.5 'What can happen when things go wrong?', within Chapter 2 of this book.

FIGURE 3.1 Cross-linking of proteins. (A) Cystine is formed by an oxidation reaction that creates a disulphide bond between two cysteine residues. (B) Lysine residues form a cross-link, with the resulting aldehydes reacting spontaneously to form an aldol

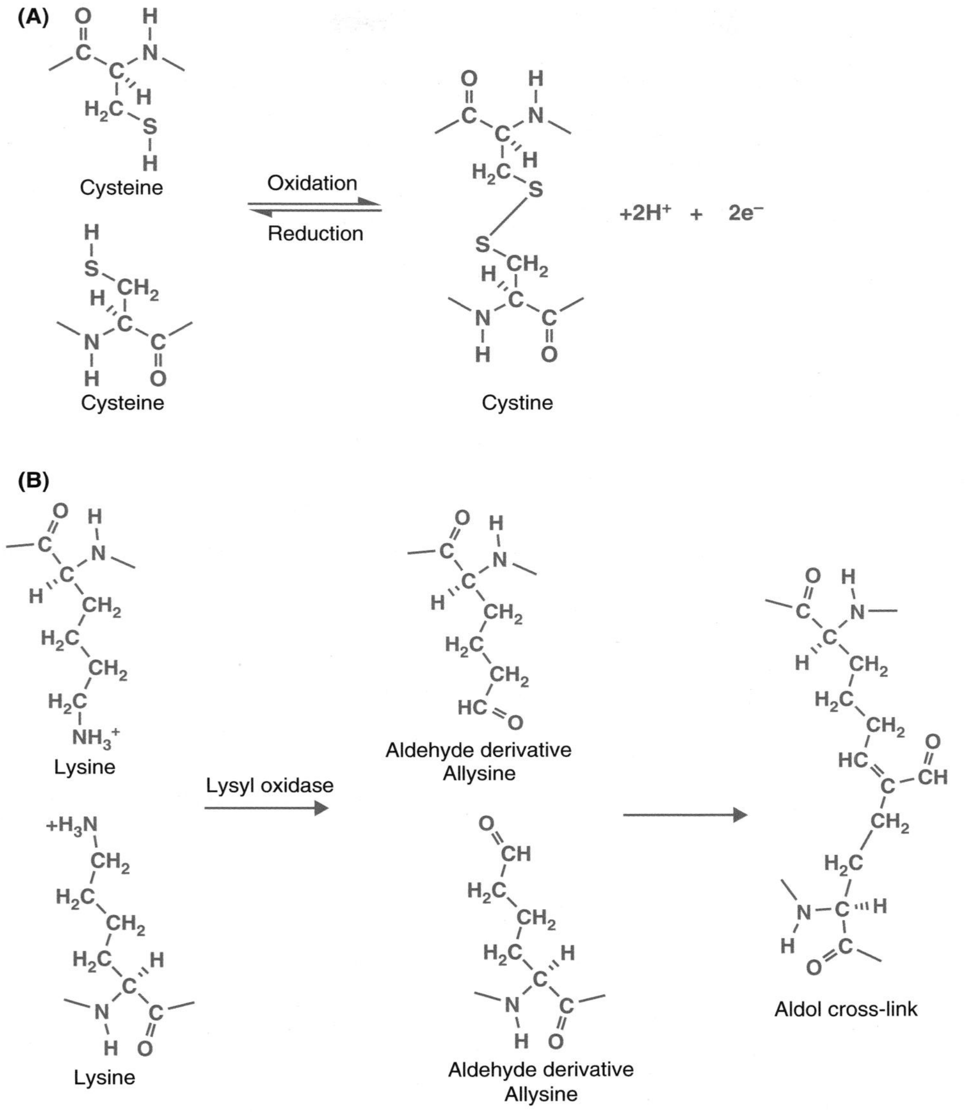

Secondary structure

The primary structure of a protein is stabilized by the formation of **hydrogen bond** interactions between adjacent amino acid residues. This subsequent arrangement is described as the *secondary structure* of a protein.

There are two common secondary structures, or *motifs*, called the *alpha (α) helix* and the *beta (β) sheet*, which are regularly repeating one-dimensional structures as shown in Figure 3.2. The stability of these structures results from the formation of hydrogen bonds between the NH and C = O groups within the polypeptide backbone.

> **KEY POINT**
>
> Polypeptide chains are stabilized by hydrogen bonds, forming different secondary protein structures.

The α helix can chemically form in either a left-handed or right-handed configuration, but the right-handed

FIGURE 3.2 Structural motifs of the α helix and β sheet within proteins, showing the position of hydrogen bond formation, indicated by dotted lines. (A) The ribbon and atomic structure of the α helix. (B) The atomic structure of an antiparallel β sheet. (C) The atomic structure of a parallel β sheet

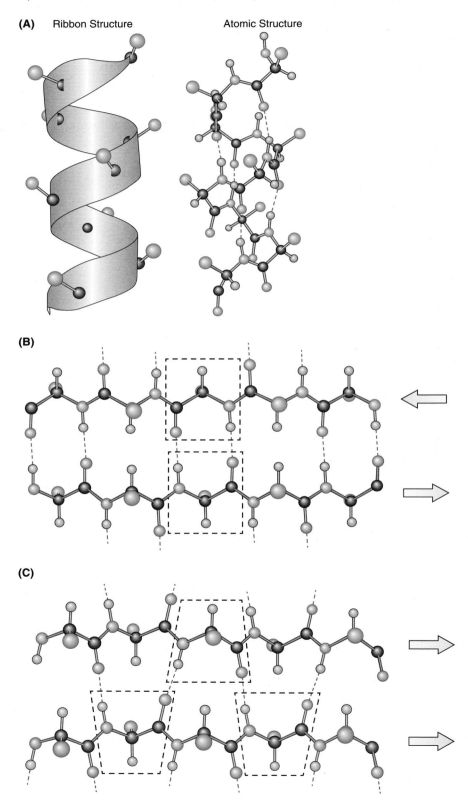

(A)   Ribbon Structure   Atomic Structure

(B)

(C)

configuration shown in figure 3.2 (A) is more common because it is energetically more favourable owing to less **steric clash** between the polypeptide backbone and the amino acid side chains. To form the α helix, the NH group of one amino acid forms a hydrogen bond with the NH group of the amino acid that is four residues ahead in the sequence. This occurs for all amino acids within the α helix except for those at the end of the chain and certain amino acids that cannot be accommodated in an α helix. The branched-chain amino acids (isoleucine, valine, and leucine) cause steric clashes, which destabilize α helices. Proline, which lacks a free amine NH_3^+ group and contains a bulky ring, deforms the α helix. The side chains of aspartate, serine, and asparagine contain hydrogen bond donors or acceptors that end up in close proximity to the main backbone and compete for the backbone NH and C = O groups, causing instability of the α-helical structure.

In the β sheet there are two or more polypeptide chains called *β strands*. As the side chains within a β strand point in opposite directions, the hydrogen bonds form with the side chains of adjacent β strands. These adjacent β strands can run in opposite directions (*antiparallel*) or in the same direction (*parallel*) as shown in Figures 3.2 (B) and (C) respectively. A single protein can contain a mixture of both parallel and antiparallel β strands within a single β sheet. In nature, the β sheets are twisted. This state is more energetically favourable, as hydrogen bonding is more likely. These structures are important in lipid **metabolism**, where the fatty acid binding proteins are almost entirely made of β sheets.

There are two other common secondary protein structures, called the *β turn* (also called the hairpin turn) and the *omega (Ω) loop*, which are often on the protein surface. These structures often contain a large number of hydrogen bonds. They are involved in molecular recognition, for example during enzyme–substrate or receptor–agonist binding, as they use their hydrogen bonds along with hydrophobic interactions to link structures.

Tertiary structure

Once the primary structure has been stabilized to form the secondary structure, the protein will then form a more compact shape and become more globular in appearance. This type of structure is called the *tertiary structure*.

This structure reduces the surface area to volume ratio of the protein, meaning there is less available space for potentially harmful molecules in the environment to interact with it. Myoglobin, the molecule responsible for carrying oxygen and binding iron within our muscles, was the first protein to have its three-dimensional structure revealed. The molecule folds so that the hydrophobic amino acid side chains are hidden in the centre of the molecule and the polar, charged side chains are on the outside. The NH and C = O groups within the hydrophobic region enable hydrogen bonding, whereas **van der Waals interactions** between the tightly packed side chains further increase stability.

Quaternary structure

The final level of protein structure is the *quaternary structure*, which describes the formation of proteins containing multiple interacting polypeptide chains called *subunits*.

> **KEY POINT**
>
> Not all proteins have the quaternary level of structure as some can exist in a monomeric state (tertiary structure).

A protein with two or more subunits is called a *multimer*. If its subunits are identical the protein is a *homo*multimer and if its subunits are different it is a *hetero*multimer. It is more economical for an organism's proteins to contain multiple identical subunits, as less DNA is required to code for it and simultaneous translation using multiple ribosomes on the same piece of DNA can occur. The simplest form of quaternary structure is a *dimer*, with two subunits. This configuration can be found in some DNA binding proteins. A protein with three subunits is a *trimer*, and one with four subunits is a *tetramer*. One example of a tetramer is haemoglobin, which has four globular protein subunits in a tetrahedral arrangement.

 Haemoglobin is covered in more detail in Section 9.2 'Erythrocytes', within Chapter 9 of this book.

> SELF CHECK 3.2
>
> How is the structure of a protein stabilized?

3.2 Enzymes and enzyme inhibition

Nearly all known enzymes are proteins (there are a few RNA-based enzymes, particularly in the ribosomes). Enzymes therefore also have a primary, secondary, and tertiary structure, and some even a quaternary structure. They are generally globular in structure, and range in size from tens to over two thousand amino acids.

The enzyme molecule, when folded, forms a three-dimensional cleft or crevice, called the *active site*, which excludes water and other non-substrate molecules. This is where the substrate binds and is where catalytic activity occurs.

The amino acids forming the active site are few in number and may be several residues apart. Residues that are adjacent to each other are more likely to be sterically prevented from interacting with each other. Residues that are further apart, however, can interact more strongly with a substrate. The predominantly non-polar environment enhances substrate binding and catalysis via the formation of multiple relatively weak interactions, including hydrogen bonds, ionic bonds, and van der Waals interactions, which would otherwise be disrupted in a polar environment. The rest of the molecule forms a protective scaffold to support the active site, and forms regulatory sites or channels to bring the substrates to the active site.

> **KEY POINT**
>
> Enzymes are proteins that have catalytic functions.

Enzymes are named according to the types of reaction that they catalyse. There are six main types of enzyme, which are classified by the Enzyme Commission (EC) as shown in Table 3.2.

Enzymes control vital biological pathways by catalysing the chemical reactions, either enhancing or reducing the speed at which they occur. The activity of enzymes is not permanent as most can be denatured, the process by which heat or chemicals cause the protein chains to unfold, destroying the configuration of amino acids in the active site. Denaturation can be reversible or irreversible, depending on the enzyme.

TABLE 3.2 Enzyme nomenclature according to the Enzyme Commission (EC)

EC number	Systematic name	Type of reaction catalysed
1	Oxidoreductases	**Oxidation–reduction** reactions
2	Transferases	Transfer of functional groups from donor to acceptor
3	Hydrolases	Hydrolysis reactions of C–O, C–N, C–C, and some other bonds, including phosphoric anhydride bonds
4	Lyases	Cleavage of C–C, C–O, C–N, and other bonds by elimination, or the addition of double bonds
5	Isomerases	Geometric or structural changes within one molecule—called isomerization reactions
6	Ligases	Joining of two molecules, coupled with the hydrolysis of a diphosphate bond in adenosine-5′-triphosphate or a similar triphosphate

The term *catalytic power* is defined as the ratio of the enzyme-catalysed speed to the uncatalysed speed of a reaction. The term *specificity* refers to the ability of the enzyme to selectively interact with its substrates. Specificity is achieved by bringing together substrates in an optimal configuration using multiple weak attractions mediated by hydrogen bonds and electrostatic and van der Waals interactions. The interaction of the enzyme with its substrate subsequently provides the level of specificity. In particular, the directional nature of hydrogen bonds helps to enforce a high level of specificity between an enzyme and its substrate. See Box 3.1 for further information on enzyme specificity.

> **KEY POINT**
>
> The most important features of enzymes are that they have catalytic power, specificity, and are highly regulated.

BOX 3.1

Enzyme specificity

There are two hypotheses for how an enzyme achieves specificity. The first is called the 'lock and key' model, proposed by Emil Fischer in the 1890s. In this model, in the same way that a key fits the shape of a specific lock in order to be able to make the mechanism open, the enzyme substrate is a complementary shape to the active site of the enzyme. However, enzymes also stabilize an intermediate part of a biological reaction called the *transition state*, making it more likely to occur. Although the lock and key model explains how specific-

ity occurs it cannot explain how stabilization occurs, and so is now thought to be incorrect. The second model is the 'induced fit' model, proposed by Daniel Koshland Jr in 1958. This model not only explains how specificity occurs but also how the transition state is stabilized. In this model, the enzyme is flexible and is constantly reshaped by interactions with the substrate. The active site can therefore bend into shape to enable a substrate to bind and is only a complementary shape once full binding occurs.

Many enzymes have a quaternary structure. Each monomer or protein chain may contain part of an active site, and so function is only achieved when multiple polypeptides interact to form the complete active site. Enzymes with multiple subunits may also have multiple active sites. These are called **allosteric** enzymes. The catalytic activity of enzymes with quaternary structure is regulated by *cooperativity*. Multimers can contain multiple active sites for a specific substrate. If one active site is occupied, there can be a subsequent change in affinity for the substrate at the other active sites. If increases in affinity occur there is *positive cooperativity*, whereas if there is a decrease in affinity there is *negative cooperativity*. As such, the quaternary structure provides a mechanism by which the enzyme subunits can communicate to alter their function.

is called a *holoenzyme*. Cofactors can be divided into two main groups: metals and small organic molecules called **coenzymes**. Coenzymes are often derived from vitamins, such as vitamin B_1 (thiamine), B_2 (riboflavin), B_3 (niacin), B_6 (pyridoxine), B_9 (folic acid), B_{12} (the cobalamins), and vitamin C, and can be bound tightly (called *prosthetic groups*) or loosely (called *cosubstrates*) to the enzyme. Enzymes using the same coenzyme will often catalyse a reaction using a similar mechanism. Examples of cofactors and the enzymes they regulate are shown in Table 3.3.

SELF CHECK 3.3

What is an enzyme active site and how is it produced?

KEY POINT

Enzyme cooperativity is a phenomenon whereby the affinity for the substrate at a active site changes in accordance to the binding of the substrate at another active site within the quaternary enzyme.

For full activity, many enzymes also need to bind to other small molecules called *cofactors*. These can be substrates or products of the reaction being catalysed or molecules further down the pathway. Binding by a cofactor causes either an increase or decrease in the enzyme activity. An enzyme in the absence of its cofactor is called an *apoenzyme* and when the cofactor binds to enable full catalytic activity the enzyme

Enzyme kinetics

Although biological reactions will take place without an enzyme present, they may occur too slowly for the demands of our bodies. Enzymes increase the speed of reactions so that essential products are formed quickly enough to keep us alive. Enzyme kinetics describe the rate at which enzymes catalyse biological reactions. The maximum rate of a reaction and the binding affinities for substrates and potential inhibitors are vital for the understanding of cellular metabolism, where each reaction is catalysed by a different enzyme. As drug molecules are often specific inhibitors targeted to particular enzymes, the pharmaceutical industry requires this information to design and create new therapeutic agents.

TABLE 3.3 Cofactors and the enzymes they regulate

Cofactor	Enzyme(s) regulated
Metal (ion)	
Copper (Cu^{2+})	Cytochrome oxidase
Iron (Fe^{2+} or Fe^{3+})	Cytochrome oxidase Catalase Peroxidase
Potassium (K^+)	Pyruvate kinase Propionyl-CoA carboxylase
Magnesium (Mg^{2+})	Hexokinase DNA polymerase Glucose-6-phosphatase
Manganese (Mn^{2+})	Arginase Superoxide dismutase
Molybdenum (Mo^{6+})	Nitrate reductase
Nickel (Ni^{2+})	Urease
Selenium	Glutathione peroxidase
Coenzyme	
Biotin	Propionyl-CoA carboxylase Pyruvate carboxylase
Coenzyme A	Acetyl-CoA carboxylase
5'-Deoxyadenosylcobalamin (vitamin B_{12})	Methylmalonyl-CoA mutase
Flavin adenine dinucleotide (FAD) (vitamin B_2)	Succinate dehydrogenase Monoamine oxidase
Nictotinamide adenine dinucleotide (NAD^+) (vitamin B_3)	Alcohol dehydrogenase Lactate dehydrogenase
Pyridoxal phosphate (vitamin B_6)	Aspartate aminotransferase Glycogen phosphorylase
Tetrahydrofolate (folic acid)	Thymidylate synthase
Thiamine pyrophosphate (vitamin B_1)	Pyruvate dehydrogenase

KEY POINT

Enzymes are catalysts and can only alter the rate of reaction, not the position of the **equilibrium** of a reaction, since the latter would be against the **laws of thermodynamics**.

It is important to appreciate that the amount of product formed stays the same, no matter whether the enzyme is present or absent. The only difference is

FIGURE 3.3 Enzymes accelerate the rate of a reaction but not the position of equilibrium. The same equilibrium point is reached by reactions in the presence or absence of an enzyme

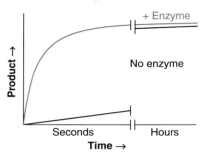

the speed at which this amount is reached. Enzymes decrease the **activation energy** required for the reaction to take place, thus the reaction can proceed at an accelerated rate. Figure 3.3 shows the rate of product formation with time when in the presence and absence of an enzyme.

When a graph of the concentration of product (on the *y*-axis) is plotted against time (on the *x*-axis), the steepness of the line shows us how quickly the product is forming. A steep line means that the product is formed more quickly than when the slope of the line is more gradual. The rate of production is measured by calculating the slope (also called the gradient) of the line at any desired time point. Figure 3.3 shows that the rate of product formation decreases over time as the gradient of the line becomes less steep as time increases. This is owing to a *steady-state equilibrium* being reached, meaning that the amount of substrate being converted into product in the forward reaction is equal to the amount of product being converted back into substrate in the reverse reaction. For each reaction, there is a corresponding rate constant (*k*). Each rate constant is dependent on the concentration of substrate and has the units time^{-1}, normally quoted in sec^{-1}.

As mentioned previously, enzymes aid the formation of a transition state. This transition state is an unstable structure that is neither the product nor the substrate. Because it is highly unstable, the transition state molecule only exists for a short amount of time. By comparison, the substrate or product molecules are relatively stable and so exist for much longer periods of time. After the reaction has taken place, the enzyme is unchanged and so it can be used repeatedly when required.

Equations to define enzyme kinetics

When a substrate binds to an enzyme, the speed or rate at which the product is formed is given the letter V for *velocity*. The rate of catalysis at the start of the reaction, V_0, is the number of **moles** of product formed per second. However, if the concentration of substrate is increased, the rate of catalysis begins to level off and approaches a limit called the *maximum rate (V_{max})*.

> **KEY POINT**
>
> An enzyme reaction will reach a *saturation point*, at which the concentration of substrate saturates the enzyme and V no longer increases.

In 1913, Leonor Michaelis and Maud Leonora Menten proposed a model to explain the concept of enzyme kinetics. Their model stated that when an enzyme E is in the presence of a substrate S, it first binds to the substrate forming an enzyme–substrate *(ES)* complex. The *ES* complex then has two fates: to dissociate back to enzyme and substrate, or for the substrate to be broken down to give a product P and release the enzyme. The reaction is reversible at each stage, and the mechanism can therefore be given the equation:

$$E + S \underset{k_{-1}}{\overset{k_1}{\rightleftarrows}} ES \underset{k_{-2}}{\overset{k_2}{\rightleftarrows}} E + P$$

Each reaction has its own rate constant (k_1 and k_2 for the forward reactions and k_{-1}, and k_{-2} for the reverse reactions). At equilibrium, the enzyme is still actively converting substrate into product, but the product is converted back to substrate at the same rate. At the start of the reaction, as close to zero as possible, there is negligible formation of product and likewise no reverse reaction. The substrate concentration ([S]) that gives a speed of half of V_{max} ($V_{max}/2$) is known as the *Michaelis constant (K_m)* and is expressed in units of concentration, usually **molar** (M). If a graph of V_0 is plotted against [S], a saturation curve is produced as shown in Figure 3.4.

The *Michaelis–Menten equation* shown below relates the rate of reaction with the concentration of substrate, which explains the data given in Figure 3.4.

$$V_0 = V_{max} \frac{[S]}{K_m + [S]}$$

FIGURE 3.4 A typical saturation curve showing how the initial reaction rate increases with substrate concentration until saturation. At saturation, the speed is maximal and is correspondingly called V_{max}

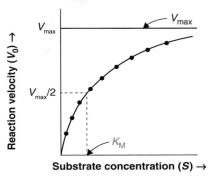

The Michaelis-Menten equation assumes that a steady state has been reached, in which the rate of formation of ES will equal the rate of breakdown of the ES complex. There are a number of other assumptions made when using this equation:

- K_m and V_{max} only define the rate of an enzyme-catalysed reaction when there is only one substrate; if there are multiple substrates then the concentration of only one can vary while the others stay constant.

- All other variables that can alter kinetics, such as temperature, pH, and ionic strength, must be constant.

- K_m and V_{max} are only valid when observing the initial rate of reaction when the product has not yet had time to be made, meaning that the concentration of product is essentially zero.

- The initial concentration of substrate must be greater than the total concentration of enzyme, which stays constant.

Once all these assumptions are agreed, the K_m and V_{max} can then be determined. However, as V_{max} is the rate of reaction at equilibrium, to measure it experimentally is impossible. Likewise, as K_m is a function of V_{max}, this too is impossible to measure. Although we now have curve-fitting software to elucidate these equations, this was not possible before computers were available. Therefore, the curved data were transformed into a straight line with the equation $y = mx + c$. There are a number of ways of doing this. If the reciprocal of both sides of the

FIGURE 3.5 A Lineweaver–Burk plot. This is a double reciprocal plot generated by plotting $1/V_0$ on the y-axis against $1/[S]$ on the x-axis. The intercept on the x-axis is $-1/K_m$ and the intercept on the y-axis is $1/V_{max}$. The gradient of the line is K_m/V_{max}

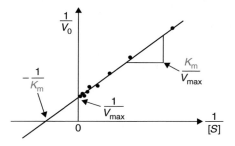

Michaelis–Menten equation is taken, this gives us the *Lineweaver–Burk* or *double reciprocal equation*:

$$\frac{1}{V_0} = \frac{K_m + [S]}{V_{max}[S]}$$

When plotted graphically with $1/V_0$ on the y-axis and $1/[S]$ on the x-axis, the line is linear and the gradient is equal to K_m/V_{max}, as shown in Figure 3.5.

This method has the inherent problem that the double reciprocal means that the errors from measurement are consequently increased. To find the intercept on the x-axis also involves a large amount of extrapolation, which is inaccurate. To combat this, other linear plots have been developed as described in Box 3.2.

BOX 3.2

Alternative enzyme kinetics plots

The Eadie–Hofstee plot, with V on the y-axis and $V/[S]$ on the x-axis, gives a straight line with the intercept of the y-axis equal to V_{max} and the gradient equal to $-K_m$. However, neither axis uses independent variables as both are dependent on V. This means that any errors in measurement will be present in both axes. In addition, the Hanes–Woolf plot with $[S]/V$ on the y-axis and $[S]$ on the x-axis gives a straight line with the intercept on the x-axis equal to $-K_m$ and the intercept on the y-axis equal to K_m/V_{max}. In this case, the gradient of the line is equal to $1/V_{max}$. Again, neither axis involves independent variables as both rely on $[S]$. None of the ways of plotting the data or the enzyme kinetic equations are perfect, but we use them accepting their inaccuracies and assumptions.

K_m and V_{max} are important as they are a measure of the effectiveness of a particular enzyme. K_m for enzymes varies greatly, from 10^{-7} M to 10^{-1} M and indicates how much substrate is needed for significant catalysis to occur. A high K_m indicates weak binding of substrate with the enzyme, whereas a low K_m indicates strong binding of substrate with the enzyme. It therefore indicates the *affinity* of the ES complex. By contrast, V_{max} provides information on the turnover of the enzyme and gives a measure of the number of substrate molecules converted into product in a unit of time, when the enzyme is fully saturated with substrate.

Allosteric enzymes, with their multiple subunits and multiple active sites, do not obey the rules of Michaelis–Menten kinetics. Here, the binding of a substrate at one active site induces changes to the molecular structure that can change the binding at other active sites. This cooperativity causes the plot of V_0 against $[S]$ to be an 'S' shape. Allosteric enzymes may also be altered by molecules that can be reversibly bound to regions other than the active sites. Their catalytic nature can therefore be adjusted to meet the needs of the cell and for this reason these enzyme are key regulators of metabolic pathways.

SELF CHECK 3.4

Why does an enzymatic reaction form a saturation curve when the reaction rate is plotted against the substrate concentration?

Enzyme inhibition

Since important biological reactions are catalysed by enzymes, it follows that many drugs have been developed to interfere with the reaction kinetics. These therapeutic agents, called *enzyme inhibitors*, may be small molecules or ions. Enzyme inhibition can be irreversible or reversible. In general, the most potent inhibitors tend to be analogues of transition state molecules. A common enzyme inhibitor that is used as a drug is aspirin, which inhibits the cyclooxygenase enzymes that produce prostaglandins during inflammation. By reversibly inhibiting these enzymes, pain and inflammation are therefore suppressed.

Section 7.5 'Eicosanoids', within Chapter 7, looks into the synthesis and function of prostaglandins.

Irreversible inhibitors and slow-binding reversible inhibitors

Irreversible enzyme inhibitors covalently modify the structure of the enzyme active site. They dissociate from the enzyme very slowly because the bonds between the enzyme and inhibitor are strong, thus preventing inhibition from being reversed. Dilution or dialysis will therefore not dislodge the inhibitor and cannot be used to reverse the effect of an irreversible inhibitor. They are normally specific for a type of enzyme. Inhibition by these molecules is not instantaneous but depends on the time spent in contact with the enzyme—the longer the inhibitor is in contact with the enzyme, the stronger the interaction between the enzyme and inhibitor.

Irreversible inhibitors bind to the enzyme as a normal substrate would, thus activating catalysis. However, an intermediate molecule is generated that covalently modifies the active site and thus irreversibly inhibits the enzyme. Examples of irreversible inhibitors are the monoamine oxidase enzyme inhibitors selegiline and tranylcypromine, which are used to treat parkinsonism and depression, respectively. Another example is the antibiotic penicillin, which covalently modifies the serine residues in the active site of the transpeptidase enzyme and so prevents bacterial cell wall synthesis, effectively killing the bacteria.

There are also some slow-binding reversible inhibitors that bind non-covalently but resemble irreversible inhibitors because their binding is so strong. One example is methotrexate, which inhibits the dihydrofolate reductase enzyme involved in folic acid metabolism and is used in the treatment of cancer and autoimmune disorders.

Reversible inhibitors

Reversible inhibitors interact with the enzyme using non-covalent interactions, including hydrogen bonds, hydrophobic interactions, and ionic bonds. The strength of inhibition is maximized by multiple weak bonds between the enzyme active site and the inhibitor. Reversible inhibitors can often be easily removed from the enzyme by dilution or dialysis, and their activity is characterized by a rapid association/dissociation with the enzyme. Reversible inhibitors are classified under four groups according to the method of inhibition:

1. *Competitive inhibition* is where the substrate and the inhibitor try to bind at the same active site and compete for the same space. Examples of clinically relevant competitive inhibitors are given in Box 3.3. This type of inhibition can be overcome by increasing the concentration of the substrate, as it will then out-compete the inhibitor for the active site. This can be shown graphically in terms of both reaction rate versus substrate concentration graphs and Lineweaver–Burk plots, as shown in Figure 3.6.

2. *Uncompetitive inhibition* is where the inhibitor binds to the ES complex, but not to the free enzyme, and forms an enzyme–substrate–inhibitor (ESI) complex. This makes the ES complex inactive and

---

BOX 3.3

Competitive inhibitors

Nitisinone is a competitive inhibitor of 4-hydroxyphenylpyruvate oxidase. It is used to treat type I tyrosinaemia, which is a disease that leads to the build-up of tyrosine in the bloodstream. Other examples of therapeutic competitive inhibitors include ibuprofen, which competitively inhibits the enzymes involved in inflammation, and statins, which competitively inhibit an enzyme in the cholesterol biosynthesis pathway to reduce blood cholesterol levels. Even some foodstuffs can act as competitive inhibitors. One example is grapefruit juice, which contains bergamottin. Bergamottin competitively inhibits the cytochrome P450 enzyme, which is involved in the metabolic breakdown of statins. If grapefruit juice is consumed at the same time as statins it can lead to the accumulation of an excessively high level of statins in the body, resulting in liver damage. Similarly, consumption of grapefruit juice with some hay fever drugs, for example terfenadine, is equally dangerous for the same reason.

FIGURE 3.6 Kinetics of a competitive inhibitor. (A) The rate of reaction (V) against substrate concentration ([S]) shows that higher concentrations of substrate are required for a certain reaction rate to be achieved by out-competing for the active site. (B) Lineweaver–Burk plot showing that K_m is increased but V_{max} is unaffected

FIGURE 3.7 Kinetics of an uncompetitive inhibitor. (A) The rate of reaction (V) against substrate concentration ([S]) shows that V_{max} is not reached, even at high concentrations of substrate. K_m decreases as more inhibitor is present. (B) Lineweaver–Burk plot showing that the gradients of the lines in the presence and absence of inhibitor are identical, as both V_{max} and K_m decrease by the corresponding amounts

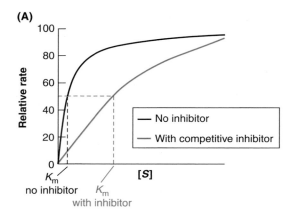

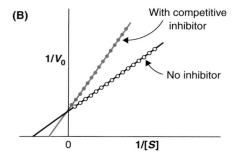

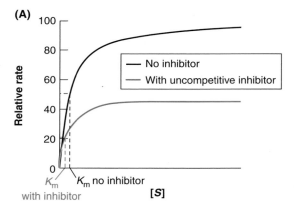

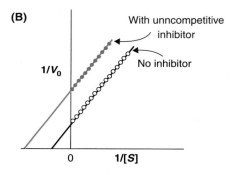

prevents the formation of product. The uncompetitive inhibitor's binding site is only formed once the ES complex occurs. In the presence of an uncompetitive inhibitor, the ESI complex cannot form product and so V_{max} is therefore reduced by comparison with the complex in the absence of inhibitor. Figure 3.7 shows the effects of an uncompetitive inhibitor. This type of inhibition is rare. Lithium, which is used as a mood stabilizer in the treatment of bipolar disorder, has been suggested to work as an uncompetitive inhibitor, although its precise mechanism of action is unknown.

3. *Mixed inhibition* involves the inhibitor binding to both the free enzyme and the ES complex, although the affinity for each is different. This is because the inhibitor and the substrate bind to the enzyme at separate sites. However, the binding of

the inhibitor affects that of the substrate, and vice versa, due to either a conformational change when one binds or by the two sites being close together, enabling interactions to occur. Increasing the substrate concentration can reduce the effects of a mixed inhibitor, but will not remove the inhibition entirely. Mixed inhibitors interfere with the substrate binding, which causes an increase in K_m, while they hinder catalysis of the ES complex, which causes a decrease in V_{max}. This type of inhibition is common and usually results from an allosteric effect. Fluoxetine, a drug that is licensed for treatment of major depression, obsessive–compulsive disorder, and severe premenstrual disorder, is thought to act as a mixed inhibitor in some circumstances and a competitive inhibitor in others.

4. *Non-competitive inhibition* can be described as one form of mixed inhibition; however, the inhibitors have identical affinities for both the free enzyme and the ES complex. The binding of the inhibitor to the enzyme again causes a reduction in the enzyme activity but without altering the ability of the substrate to bind. The extent of the inhibition therefore relies only on the concentration of the inhibitor. One example of a non-competitive inhibitor is donepezil, which is a cholinesterase inhibitor used in the treatment of Alzheimer's disease. People with this disease have a low level of acetylcholine and so the enzyme that breaks down the acetylcholine in the brain must be inhibited to help maintain a functional level of acetylcholine. A further example is the antiretroviral drug nevirapine, which is a non-nucleoside reverse transcriptase inhibitor used in the treatment of human immunodeficiency virus and acquired immunodeficiency syndrome. Figure 3.8 shows the effects of a non-competitive inhibitor.

SELF CHECK 3.5

Irreversible inhibitors can be used to map enzyme active sites, but how?

FIGURE 3.8 Kinetics of a non-competitive inhibitor. (A) The rate of reaction (V) against substrate concentration ($[S]$) shows that V_{max} is not reached, as the inhibitor binds to both the free enzyme and the enzyme–substrate complex. K_m is unchanged but the reaction rate increases more slowly at low concentrations of substrate. (B) Lineweaver-Burk plot showing that K_m is unaffected but V_{max} is reduced

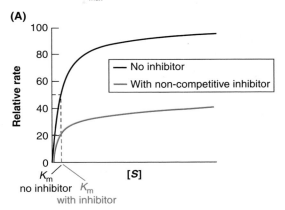

(A)

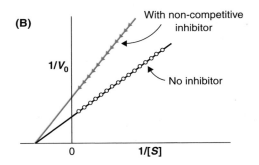

(B)

3.3 Metabolic pathways and abnormal metabolism

Metabolism is the conversion of nutrients into energy as well as the use of energy and nutrients to build cellular components. It occurs in several series of enzyme-controlled reactions within cells called metabolic pathways. Enzymes involved in metabolic pathways may exist individually or form complexes in which the substrate is sequentially modified and passed to the next enzyme along. At each stage, intermediates and reaction products are formed. Formation of enzyme complexes prevents the loss of intermediates and ensures progression to the next stage of the pathway. Enzymes can also be bound together within the cell membrane, which constrains the movement of both enzyme and substrate.

Metabolism occurs in two forms: anabolism and catabolism. Anabolism requires energy (is *endergonic*) to synthesize cellular components such as nucleic acids, polysaccharides, lipids, or proteins from their precursors (nucleotides, sugars, fatty acids, and amino acids, respectively). The energy helps form covalent bonds, which join the precursors together. By contrast, catabolism breaks down carbohydrates, lipids, and proteins to form breakdown products and release energy (is *exergonic*). The anabolic and catabolic reactions are linked, as the products made by one type of reaction are often used as substrates for the other type of reaction. They act simultaneously within the cell, but are strictly and separately regulated. Interference and competition are prevented by compartmentalizing the opposing reactions into separate cellular regions. Metabolic fuel sources come from our dietary intake of carbohydrates, fats or lipids, and proteins.

> **KEY POINT**
>
> Metabolism involves a series of reactions within organisms that enable life.

Energy molecules and activated carriers

There are a number of energy molecules within the cell, each with a different role; however, adenosine-5′-triphosphate (ATP) is the principal energy currency in cells. ATP hydrolysis enables the chemical energy from the high-energy phosphate bonds within the molecule to be released. The products are adenosine-5′-

diphosphate (ADP) and inorganic phosphate. ADP can be further hydrolysed to adenosine-5′-monophosphate (AMP) with the removal of a further molecule of inorganic phosphate. Carbohydrates and lipids are good sources of energy due to the reduced state of their carbon atoms, and they release *hydride ions* (a proton with two electrons) when degraded. These hydride ions are transferred to *activated carriers*:

- *Nicotinamide adenine dinucleotide (NAD⁺)* is reduced to NADH in the equation shown in Figure 3.9 (A), in which hydride ions are transferred to NAD⁺ by *dehydrogenase enzymes*. The NADH is finally oxidized back to NAD⁺ by releasing its reducing equivalents to acceptor systems within mitochondria. These acceptor systems or oxidizing agents include oxygen molecules (O_2), which are subsequently reduced to water (H_2O). The energy released is coupled to the formation of ATP in the process of *oxidative phosphorylation*.

- *Flavin adenine dinucleotide (FAD)* is reduced to FADH$_2$ as it can accept two electrons. However, unlike NAD⁺ it also takes up two protons, as shown in Figure 3.9 (B).

FIGURE 3.9 Oxidation–reduction reactions of (A) nicotinamide adenine dinucleotide (NAD⁺) and (B) flavin adenine dinucleotide (FAD)

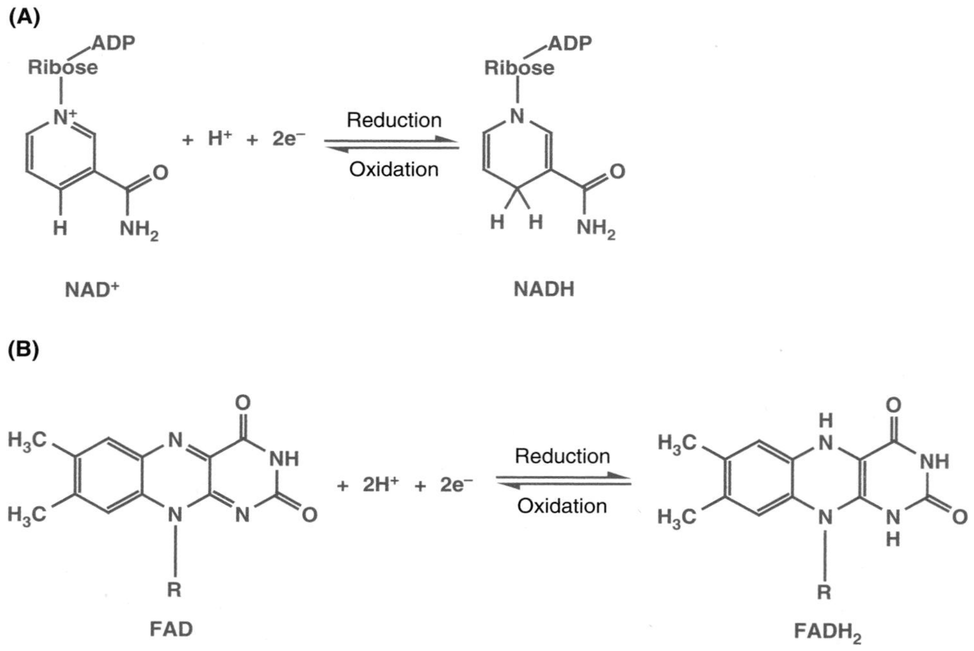

- *Nicotinamide adenine dinucleotide phosphate (NADP⁺)* forms NADPH when it is reduced with electrons in the form of hydride ions. The electrons are then removed from NADPH by NADP⁺-specific dehydrogenases.

- *Coenzyme A (CoA)* carries acyl groups that are important in both fatty acid catabolism and membrane lipid synthesis (anabolism). Acyl groups are linked to CoA by thioester bonds to form acetyl-CoA.

> **KEY POINT**
>
> Activated carriers are kinetically stable molecules and so enzymes are required to control the flow of reducing power and energy. The relatively small number of activated carriers helps to simplify metabolism.

Carbohydrates

Carbohydrates are the main fuel source in our bodies and are divided into two groups: simple sugars and complex carbohydrates. Simple sugars like glucose are metabolized directly via *glycolysis* and the *citric acid cycle*, whereas complex carbohydrates like starch and glycogen are first broken down into simple sugars. The process of glycogen breakdown is called *glycogenolysis*.

If the intake of carbohydrates exceeds the amount the body needs, the excess is converted into triacylglycerols and glycogen for long-term storage. However, if carbohydrate intake is low, for example during fasting, fatty acid breakdown in the liver or kidney occurs. Fatty acids and triacylglycerols are the subunits of lipids, which will be considered under 'Fatty acids' within this section.

Glycolysis

Glycolysis is a catabolic process that occurs in the cytoplasm, by which glucose is degraded to produce energy. It can occur in the absence of oxygen (anaerobic conditions). Glycolysis can be separated into two stages, as shown in Figure 3.10. The first stage converts glucose into two molecules of glyceraldehyde 3-phosphate. In the second stage, the glyceraldehyde 3-phosphate is oxidized to form pyruvate. It is during this second stage that the energy from the glucose molecule is extracted.

> **KEY POINT**
>
> Glycolysis is the first part of the metabolic pathway during which glucose and other simple sugars are catabolized and energy is produced.

The ten reactions that occur during glycolysis are described below. The first stage encompasses reactions 1 to 5, while the second stage covers reactions 6 to 10.

1. *Phosphorylation of glucose at the 6-carbon position to form glucose 6-phosphate.* To prevent glucose diffusing out of the cell it is phosphorylated. This irreversible reaction, catalysed by hexokinase, requires energy and so one ATP molecule is broken down to ADP. This enzyme, as per all kinases, requires Mg^{2+}, and is allosterically inhibited by the product, glucose 6-phosphate. This stage is highly regulated and inhibition is only removed by subsequent reactions consuming the glucose 6-phosphate. In liver cells this reaction is catalysed by the enzyme glucokinase, not hexokinase.

2. *Formation of fructose 6-phosphate.* This is a reversible isomerization step, catalysed by glucose-6-phophate isomerase, in which an aldose, glucose 6-phosphate, is converted to a ketose, fructose 6-phosphate. The glucose 6-phosphate is initially a cyclic molecule and so the ring is first opened before the isomerization can take place. The ring then reforms to give fructose 6-phosphate.

3. *Phosphorylation of fructose to form fructose 1,6-bisphosphate.* The prefix 'bis' means that the two phosphate groups are on separate carbon atoms. This irreversible reaction again requires Mg^{2+} and utilizes another molecule of ATP. It is catalysed by phosphofructokinase, which is an allosteric enzyme that determines the speed of glycolysis. The reaction ensures that the cell metabolizes glucose rather than allowing its conversion to another sugar for storage. It is therefore highly regulated.

4. *Cleavage of fructose 1,6-bisphosphate to form two triose phosphates.* This reaction, catalysed by fructose-bisphosphate aldolase (often just

FIGURE 3.10   The glycolytic pathway

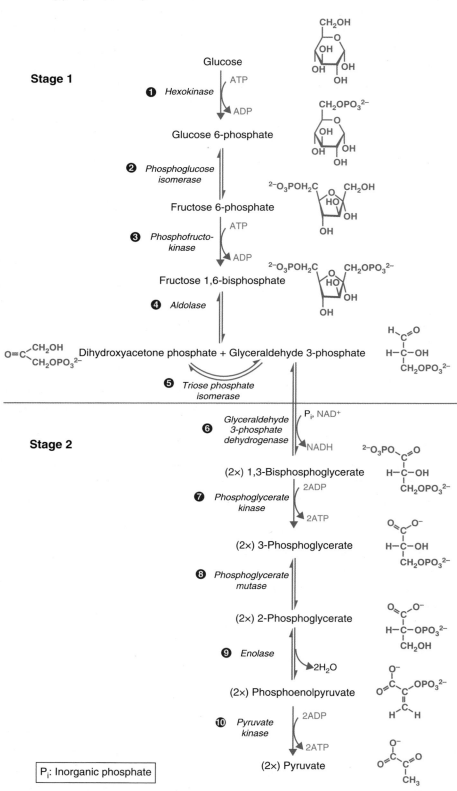

Stage 1

Glucose

❶ *Hexokinase*   ATP → ADP

Glucose 6-phosphate

❷ *Phosphoglucose isomerase*

Fructose 6-phosphate

❸ *Phosphofructo-kinase*   ATP → ADP

Fructose 1,6-bisphosphate

❹ *Aldolase*

Dihydroxyacetone phosphate + Glyceraldehyde 3-phosphate

❺ *Triose phosphate isomerase*

Stage 2

❻ *Glyceraldehyde 3-phosphate dehydrogenase* P_i, NAD⁺ → NADH

(2×) 1,3-Bisphosphoglycerate

❼ *Phosphoglycerate kinase*   2ADP → 2ATP

(2×) 3-Phosphoglycerate

❽ *Phosphoglycerate mutase*

(2×) 2-Phosphoglycerate

❾ *Enolase*   → 2H₂O

(2×) Phosphoenolpyruvate

❿ *Pyruvate kinase*   2ADP → 2ATP

(2×) Pyruvate

P_i: Inorganic phosphate

aldolase), gives rise to two triose phosphate products—glyceraldehyde 3-phosphate and dihydroxyacetone phosphate. While glyceraldehyde 3-phosphate is on the direct pathway of glycolysis, dihydroxyacetone phosphate is not. It is, however, converted to glyceraldehyde 3-phosphate during the next step and thus re-enters glycolysis.

5. *Interconversion of glyceraldehyde 3-phosphate and dihydroxyacetone phosphate.* This rapid and reversible reaction is catalysed by triosephosphate isomerase. At equilibrium, 96% of the triose phosphate is dihydroxyacetone phosphate; interconversion is triggered as the further stages of glycolysis reduce the amount of glyceraldehyde 3-phosphate. This reaction completes the first stage of glycolysis.

6. *Conversion of glyceraldehyde 3-phosphate into 1,3-bisphosphoglycerate.* This reaction involves two steps. The aldehyde group is first oxidized to a carboxylic acid by NAD^+, and second, the carboxylic acid is then joined to an orthophosphate (PO_4^{3-}). The NAD^+ is reduced to NADH. This reaction is catalysed by glyceraldehyde-3-phosphate dehydrogenase.

7. *Transfer of a phosphoryl group from 1,3-bisphosphoglycerate to ADP to form ATP and 3-phosphoglycerate.* This reaction, catalysed by phosphoglycerate kinase, is known as a substrate-level phosphorylation and requires a Mg^{2+} ion. As each glucose molecule gives two molecules of glyceraldehyde 3-phosphate, there are therefore two molecules of ATP produced at this stage. With the production of ATP during this reaction, the energy debt from the first stage of glycolysis is paid off.

8. *Movement of the phosphoryl group from 3-phosphoglycerate to form 2-phosphoglycerate.* Phosphoglycerate mutase catalyses this reaction and produces a substrate molecule that can be further transformed during the next reaction.

9. *Dehydration of 2-phosphoglycerate.* This reaction is catalysed by enolase, and introduces a double bond to create an enol called phosphoenolpyruvate, which is an unstable molecule and is further transformed during the next reaction.

10. *The conversion of phosphoenolpyruvate to pyruvate, generating ATP.* Phosphoenolpyruvate quickly releases its phosphoryl group to a further molecule of ADP, forming ATP and a molecule of pyruvate. Pyruvate is a more stable molecule than phosphoenolpyruvate. This final reaction is catalysed by pyruvate kinase and regulated by the concentration of ATP. Again, as there are two molecules of glyceraldehyde 3-phosphate entering stage two of glycolysis, there are two molecules of pyruvate and a further two molecules of ATP formed by the end of glycolysis.

> **KEY POINT**
>
> The overall net reaction of glycolysis is:
>
> Glucose + 2ADP + 2P$_i$ + 2NAD$^+$ $\rightarrow$ 2Pyruvate + 2ATP + 2NADH + 2H$^+$ + 2H$_2$O
>
> where P$_i$ denotes inorganic phosphate.

At the end of glycolysis, the important process is to prevent a lack of NAD^+ from becoming limiting. Pyruvate is therefore further metabolized so that NAD^+ is regenerated from NADH. Under anaerobic conditions, for example in contracting muscle cells during intense activity, pyruvate can be reduced by NADH to form lactate (also known as lactic acid—the substance known to cause muscle cramps after intense exercise) and NAD^+. However, under aerobic conditions pyruvate is further metabolized in the *citric acid cycle* (also known as the tricarboxylic acid cycle or Krebs cycle).

Citric acid cycle

The pyruvate generated during glycolysis is first transported from the cytoplasm into the mitochondria by a specific carrier protein. Once in the mitochondrial matrix, the pyruvate is oxidatively decarboxylated by pyruvate dehydrogenase to form acetyl coenzyme A (acetyl-CoA). This produces carbon dioxide (CO_2) and a molecule of NADH. The NADH is then reoxidized in the mitochondrial electron transport chain. Pyruvate dehydrogenase is a multienzyme complex that is a non-covalent group of three enzymes and five coenzymes. The acetyl-CoA formed is the fuel for the citric acid cycle, which is a series of eight oxidation–reduction reactions, as shown in Figure 3.11.

FIGURE 3.11 The citric acid cycle

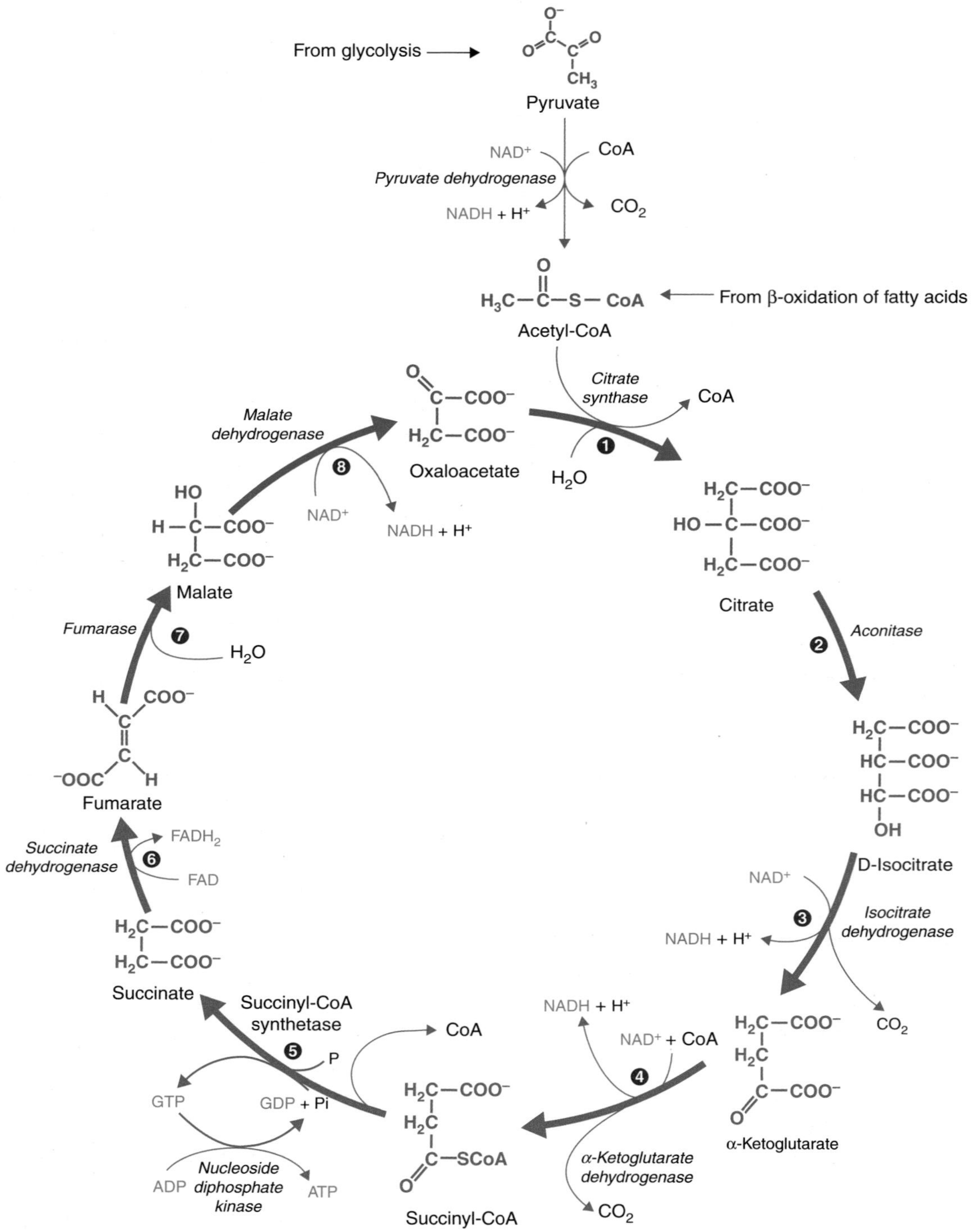

The citric acid cycle occurs in the mitochondria and is the final pathway for the oxidation of the fuel molecules.

The eight reactions occurring in the citric acid cycle are described below:

1. *Condensation of acetyl-CoA with oxaloacetate.* This produces citrate and is catalysed by citrate synthase. This enzyme is a dimer that is conformationally changed when oxaloacetate binds to the active site. This change in shape aids the binding of acetyl-CoA and closes the active site to stop water from entering. This prevents the hydrolysis of acetyl-CoA to acetate, which would be detrimental to the citric acid cycle. Both succinyl-CoA, from later in the citric acid cycle, and NADH are allosteric inhibitors of citrate synthase and thus exert some control of the citric acid cycle.

2. *Citrate isomerization into isocitrate.* This reaction is catalysed by aconitate hydratase (known as aconitase) and the isocitrate produced is the start of four oxidation–reduction reactions.

3. *Isocitrate is oxidatively decarboxylated to α-ketoglutarate, with the concomitant reduction of NAD+ to NADH.* The rate of α-ketoglutarate formation, catalysed by isocitrate dehydrogenase, determines the overall rate of the citric acid cycle.

4. *The α-ketoglutarate is oxidatively decarboxylated to form succinyl-CoA, with the further concomitant reduction of NAD+ to NADH.* This reaction is catalysed by 2-oxoglutarate dehydrogenase (α-ketoglutarate dehydrogenase), which again is a multienzyme complex with four enzymes and five coenzymes. Succinyl-CoA is a high-energy intermediate used to drive the phosphorylation of guanosine-5′-diphosphate (GDP) to guanosine-5′-triphosphate (GTP).

5. *Cleavage of the thioester bond within succinyl-CoA to form succinate.* The reaction is catalysed by succinyl-CoA synthetase (also known as succinyl-CoA ligase) and forms a molecule of GTP. This GTP is interconverted to ATP by nucleoside diphosphokinase, whereby the terminal phosphate from the GTP is donated to ADP to form ATP in the following reaction:

$$GTP + ADP \leftrightarrow ATP + GDP$$

6. *Oxidation of succinate to fumarate.* This oxidation involves the removal of hydrogen atoms. The enzyme involved is succinate dehydrogenase, which is a membrane-bound enzyme that is part of the mitochondrial electron transport chain. This reaction is exergonic enough to reduce FAD to $FADH_2$. Succinate dehydrogenase is a dimer and the FAD covalently binds to one of the subunits.

7. *Fumarate is hydrated to form L-malate.* The hydration step, catalysed by fumarate hydratase (fumarase), converts fumarate into an oxidizable product (malate), enabling further conversion in the cycle.

8. *Oxidation of malate to oxaloacetate.* Malate dehydrogenase catalyses this final reaction in the citric acid cycle, which is coupled to the reduction of a final molecule of NAD+. The concentration of oxaloacetate within the mitochondrial matrix is usually low, as the reaction involving citrate synthase (the first reaction within the citric acid cycle) is more favourable and quickly removes the final product.

The overall net equation of the citric acid cycle is:

Acetyl-CoA + 3NAD+ + FAD + ADP + P_i + $2H_2O \rightarrow$ CoA + $2CO_2$ + 3NADH + $FADH_2$ + ATP + 2H+

where P_i denotes inorganic phosphate.

When glycolysis and the citric acid cycle are taken together, the net equation is:

Glucose + $2H_2O$ + 10NAD+ + 2FAD + 4ADP + $4P_i \rightarrow$ $6CO_2$ + 10NADH + 10H+ + $2FADH_2$ + 4ATP

The 12 reduced coenzymes (10 NADH and 2 $FADH_2$) generated from one molecule of glucose which has undergone glycolysis and the citric acid cycle then produce a further maximum of 34 ATP molecules in the electron transport and oxidative phosphorylation pathways, since each NADH molecule can yield 3 ATP molecules, and each $FADH_2$ can yield 2 ATP molecules. Box 3.4 emphasizes the importance of the pathways involving carbohydrates.

Diseases of carbohydrate metabolism

As change to the key stages within the citric acid cycle would be severely detrimental, diseases involving them are rarely seen as they are more than likely lethal to the individual. However, diseases due to altered glycolysis can occur. In particular, red blood cells can be severely affected as they have no mitochondria and so cannot obtain energy through the citric acid cycle. Deficiency of one or more enzymes in the glycolysis pathway within red blood cells results in their destruction, as their energy needs are not met. This causes haemolytic anaemia, the most common cause of which is pyruvate kinase deficiency, which leads to the accumulation of 3-phosphoglycerate.

 For further information on red blood cells and anaemias see Section 9.2 'Erythrocytes' and Section 9.5 'Blood disorders', in Chapter 9 of this book.

McArdle's syndrome is an autosomal recessive genetic disorder in which a defect in the glycogen phosphorylase gene results in an inability to break down glycogen in the muscle. People with McArdle's syndrome cannot intensively or excessively exercise as they cannot break down glycogen to glucose. Instead, glucose 1-phosphate is formed and, since this molecule has a phosphate group attached, it means that it cannot leave the cell via the glucose transporters. Muscle activity is then dependent on the availability of glucose circulating in the bloodstream rather than from storage.

Gluconeogenesis

Both the brain and red blood cells use glucose as their only fuel. This leads to the question of what happens if the supply of glucose runs out. The answer is that there are mechanisms in place to generate glucose from other substances. One of these processes is called gluconeogenesis and it occurs in the liver or kidney during starvation or fasting. This pathway converts pyruvate into glucose, as shown in Figure 3.12.

In addition to pyruvate, three other fuels can also enter the gluconeogenesis pathway. These are lactate from skeletal muscle, glycerol from storage, and amino acids from the diet or from the breakdown of muscle. The breakdown of muscle occurs particularly during periods of starvation; some of the amino acids released enter the gluconeogenesis pathway via oxaloacetate or pyruvate. Lactate is converted to pyruvate using lactate dehydrogenase, and the pyruvate then enters gluconeogenesis. The hydrolysis of triacylglycerols in the adipose tissue forms glycerol and fatty acids. The glycerol then enters the gluconeogenesis pathway via dihydroxyacetone phosphate.

Gluconeogenesis is not simply the reverse of glycolysis, as some of the steps are irreversible and involve the use of different enzymes, as detailed in Figure 3.12. The first step, conversion of pyruvate to oxaloacetate, takes place inside the mitochondria but the rest of the pathway takes place in the cytoplasm. The oxaloacetate is then reduced to malate by an NADH-linked malate dehydrogenase inside the mitochondria. Malate can cross the mitochondrial membrane and once in the cytoplasm it is reoxidized to oxaloacetate by NAD^+-linked malate dehydrogenase. The next stages then take place in the cytoplasm, until the final step, where glucose is formed from glucose 6-phosphate. This final step takes place inside the lumen of the endoplasmic reticulum of liver or kidney cells only. In other tissues, gluconeogenesis finishes with glucose 6-phosphate, which cannot be transported out of the cell and so accumulates to be stored as glycogen in the process of *glycogenesis*.

KEY POINT

Gluconeogenesis is the formation of glucose from pyruvate and should not be confused with glycogenesis, which is the formation of glycogen from glucose.

SELF CHECK 3.6

What would happen to the metabolism of a long-distance runner if they did not stock up on carbohydrates, and why?

FIGURE 3.12   Gluconeogenesis and the entry points for alternative fuels

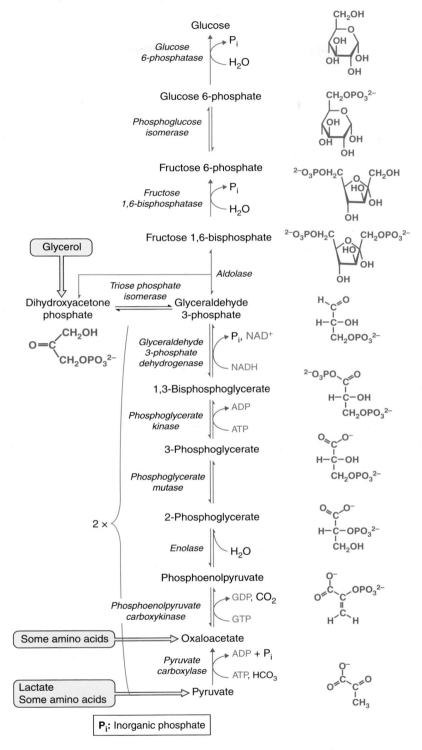

Regulation of carbohydrate metabolism

Carbohydrate metabolism is fundamental in generating energy for biological processes under normal circumstances. Controlling the blood glucose level is important for the functioning of many organs and it is unsurprising therefore that there are numerous regulatory stages in carbohydrate metabolism.

Phosphofructokinase is the most important controlling enzyme in glycolysis and is inhibited by high levels of ATP. It has two ATP binding sites. The first is a high-affinity catalytic site and the second is a low-affinity regulatory site. When ATP binds to the catalytic site the enzyme's affinity for its other substrate, fructose 6-phosphate, is reduced. AMP prevents permanent inhibition of phosphofructokinase by competing with ATP for the allosteric regulatory site. As the concentration of ATP falls due to consumption, the ratio of ATP to AMP is reduced and this stimulates glycolysis. In muscle, a decrease in pH also inhibits phosphofructokinase. The pH falls due to the formation of lactic acid during intense activity that induces anaerobic conditions (for example a sprint). Inhibition prevents muscle damage by accumulation of too high a concentration of lactic acid. By contrast, in aerobic conditions such as a long, slow run, the citric acid cycle is activated and the pyruvate is converted to CO_2 and H_2O rather than lactate. In the liver, lactate is not normally formed and so low pH is not regulatory here. Instead, phosphofructokinase is inhibited by citrate, a metabolite formed in the citric acid cycle, which enhances the inhibitory effect of ATP. A high level of cytoplasmic citrate ensures that there are sufficient metabolic precursors, meaning further glycolysis is unnecessary.

When phosphofructokinase is inactive, the level of fructose 6-phosphate is elevated and this elevates the concentration of glucose 6-phosphate as the two products are in equilibrium. The glucose 6-phosphate can then be converted to glycogen for storage.

This is not the only glycolysis step to have physiological regulation; the pathway is also regulated at the first step, whereby hexokinase is inhibited by its product, glucose 6-phosphate. In the muscle, accumulation of glucose 6-phosphate signals that glucose is no longer needed for energy and it therefore remains in the bloodstream and is subsequently stored as glycogen (in the liver) or used to synthesize fatty acids.

While hexokinase is involved in most cells, glucokinase is used in the liver when blood glucose levels are high. Glucokinase has approximately 50-fold lower affinity for glucose than hexokinase. This lower affinity ensures that when glucose levels are low, the glucose is used up by the brain and muscles first. This is also a point at which positive feedback can occur, as outlined in Integration Box 3.1.

The final reaction within glycolysis is also regulated. There are several isoforms of the enzyme pyruvate kinase: the muscle contains type M, whereas the liver contains type L. ATP allosterically inhibits pyruvate kinase, and when concentrations of ATP are high, the rate of glycolysis is reduced. This allosteric inhibition is also achieved by the amino acid alanine, which can be synthesized from and broken down to pyruvate, as described under 'Proteins' within this section. When the level of ATP is low, fructose 1,6-bisphosphate from earlier in the pathway activates pyruvate kinase to enhance the rate of glycolysis. In the liver, there is an additional controlling feature via reversible phosphorylation. When blood glucose is low, the pancreatic hormone **glucagon** triggers a messenger

> ### INTEGRATION BOX 3.1
>
> ## Positive feedback
>
> Glucokinase is also present in the β-cells of the pancreas. The resulting formation of glucose 6-phosphate in this organ leads to the secretion of insulin, the hormone involved in stimulating the uptake of glucose by the cells. This type of positive feedback signals for glucose to be removed from the bloodstream. The glucose is then metabolized by glucokinase to form more glucose 6-phosphate, which is the precursor for glycogen formation.
>
> You can read more about insulin in Section 7.6 'Hormones', within Chapter 7 of this book. Section 8.2 'Feedback mechanisms to achieve homeostasis', within Chapter 8, explains positive feedback systems.

cascade involving **3′,5′-cyclic adenosine monophosphate (cAMP)**. This phosphorylates the pyruvate kinase type L and thus reduces its activity to prevent glucose consumption by the liver when the brain's need is greater.

Fatty acids

The triacylglycerols from the diet are partially degraded by lipases in the stomach, but in the duodenum, the pancreatic enzymes, including pancreatic lipase and esterases, raise the pH and hydrolyse the fatty acid ester bonds. The addition of **bile salts** helps to emulsify the triacylglycerols and aid pancreatic enzyme activity. The resulting short-chain fatty acids, which have fewer than six carbon atoms, are absorbed from the intestine and subsequently transported to the mitochondria, where they are broken down through a process called *β-oxidation*. Long-chain fatty acids, which have more than six carbon atoms, are first condensed with glycerol to form alternative triacylglycerols. These then enter the **lymphatic system**, and subsequently the bloodstream, and are transported to other organs where they are further catabolized.

 You can learn more about the chemistry of triacylglycerols, fatty acids, and lipids in Chapter 9 'Chemistry of biologically important molecules', within the *Pharmaceutical Chemistry* book of this series.

The fatty acids are first linked to coenzyme A (CoA) via a thiol ester bond. This reaction is catalysed by acyl-CoA synthetase (also called fatty acid thiokinase) in the mitochondria or at the endoplasmic reticulum surface. It requires energy and so one molecule of ATP is broken down to AMP.

For **saturated** fatty acids, β-oxidation occurs in a cycle of events as shown in Figure 3.13. One cycle of these reactions results in a fatty acid that has been shortened by two carbon atoms and one molecule of acetyl-CoA. The cycle is repeated several times until two molecules of acetyl-CoA remain. For **unsaturated** fatty acids, an isomerase and reductase are required to first modify the double bond.

> **KEY POINT**
>
> Fatty acid breakdown (β-oxidation) is a four-step cycle which is repeated until the long-chain fatty acid is broken down to numerous acetyl-CoA products.

FIGURE 3.13 β-Oxidation of saturated fatty acids

Fatty acyl-CoA

$R-CH_2-CH_2-CH_2-C\overset{O}{\underset{SCoA}{<}}$

Shortened fatty acyl-CoA

$R-CH_2-C\overset{O}{\underset{SCoA}{<}}$

Repeat cycle...

3-Ketoacyl-CoA thiolase

Cleavage

$CH_3-C\overset{O}{\underset{SCoA}{<}}$
Acetyl CoA

FAD
Acyl-CoA dehydrogenase
FADH₂

Oxidation

$R-CH_2-CH=CH-C\overset{O}{\underset{SCoA}{<}}$ 2,3-Enoyl-CoA

Hydration H_2O *Enoyl CoA hydratase*

CoA

$R-CH_2-\overset{O}{\overset{||}{C}}-CH_2-C\overset{O}{\underset{SCoA}{<}}$
3-Ketoacyl-CoA

$R-CH_2-CHOH-CH_2-C\overset{O}{\underset{SCoA}{<}}$

L-3-Hydroxyacyl-CoA

Oxidation

NADH + H⁺   NAD⁺

3-Hydroxyacyl-CoA dehydrogenase

The acetyl-CoA from each stage enters the citric acid cycle if there is a balance between fat and carbohydrates; there must be sufficient oxaloacetate from carbohydrate catabolism to combine with the acetyl-CoA. Under fasting conditions or diseases such as diabetes, the oxaloacetate is used to produce glucose by the gluconeogenic pathway. The acetyl-CoA cannot then enter the citric acid cycle as there is no oxaloacetate and instead forms acetoacetate and D-3-hydroxybutyrate. These products are collectively known as *ketone bodies*. The acetoacetate is slowly decarboxylated to acetone, which can be detected in the breath and urine of a diabetic person. Case Study 3.1 considers the link between diabetes and the formation of ketone bodies in the urine.

Not only are lipids essential components of cell membrane phospholipids, they can also be used as energy sources. If consumed in excess of usage, the surplus is stored as triglycerides within the adipose tissue. However, some can be deposited within the vessels of the cardiovascular system and can cause atherosclerosis and heart disease. Excess fat is also correlated with an increased risk of some cancers, including breast and colon cancers. Conversely, if intake of lipids is low, there is a risk of deficiency diseases, as some fatty acids are *essential*, that is they cannot be made within the body. These include linoleic acid, linolenic acid, and arachidonic acid.

CASE STUDY 3.1

Huri is a type I diabetic and has been on insulin all of his life. He has gone to the GP after feeling very tired and nauseous and having passed out on a number of occasions. The GP took a routine blood and urine sample and told him he has high levels of ketones in his urine. The GP explained to Huri that he has developed ketoacidosis because of his diabetes.

REFLECTION QUESTIONS

1. What is diabetes and what are the different types of diabetes?

2. Why do ketone bodies form in diabetic patients?

3. What are the symptoms of ketoacidosis?

Answers

1 Diabetes is a group of metabolic diseases characterized by a high blood glucose level (hyperglycaemia). In the short-term it leads to ketoacidosis. If left untreated it leads to long-term complications including damage to the retina in the eye, nerve damage, and kidney damage. There are two main types of diabetes called type I and type II. Type I is where there is an insulin deficiency or production is impaired. Type II is where there may be only partial secretion of insulin or the cells are insulin resistant. There is also a third type of diabetes called gestational diabetes that is brought on by pregnancy. This type of diabetes is due to insufficient secretion of insulin.

2 Ketones are formed when carbohydrate intake is low and the body uses fat as an energy source. The acetyl coenzyme A from fatty acid catabolism normally enters the citric acid cycle as long as there is a balance of fat and carbohydrates. This means that there must be sufficient oxaloacetate from carbohydrates to combine with the acetyl coenzyme A. However, in diabetic patients, the oxaloacetate is used up to produce glucose by the gluconeogenic pathway. The acetyl coenzyme A cannot then enter the citric acid cycle as there is no oxaloacetate and instead forms keto-acids (also known as ketone bodies) including acetoacetate and D-3-hydroxybutyrate which are excreted through the urine (ketonuria).

3 In unmanaged diabetic patients, glucose levels in the blood are high, but cells do not absorb it, which means they are starved. The body increases fat and protein breakdown to cope. Byproducts of this are keto-acids. These enter the bloodstream and cause the blood to become acidic. This disrupts the homeostasis of electrolytes in the body (potassium is excreted in the urine) and this can have serious effects on the body. There are various symptoms that develop as a consequence of excessive ketones in the blood, including: dehydration, dry skin, fruity smelling breath, vomiting, arrhythmia, fatigue, loss of consciousness and in severe cases it may lead to a coma.

Proteins

Amino acids from proteins can be converted into glucose, which is then metabolized to give energy or converted to fatty acids. If the dietary levels of lipids and carbohydrates are sufficient for the energy needs of the organism, excess amino acids are converted to triacylglycerols and stored in the adipose tissue.

Amino acid catabolism usually takes place in the liver, although some can occur in the muscle. Firstly, the amino group is removed using transaminase enzymes and transferred to α-ketoglutarate to form glutamate. The α-ketoglutarate is then oxidatively deaminated to form the ammonium cation (NH_4^+). The NH_4^+ in our bodies is converted to urea for excretion. Once this is completed, the remaining carbon skeleton is metabolized by the formation of glucose, a citric acid intermediate, or acetyl-CoA.

Figure 3.14 shows that all amino acids are broken down to one of the following products: acetyl-CoA, acetoacetyl-CoA, fumarate, α-ketoglutarate, oxaloacetate, pyruvate, or succinyl-CoA. Those forming fumarate, α-ketoglutarate, oxaloacetate, pyruvate, or succinyl-CoA are called *glucogenic amino acids* and those forming acetyl-CoA or acetoacetyl-CoA are called *ketogenic amino acids*.

Diseases such as alkaptonuria and phenylketonuria are genetic disorders that are brought about by faulty amino acid metabolism. Alkaptonuria occurs due to a defect in an enzyme called homogentisate 1,2-dioxygenase, which is involved in tyrosine metabolism. A toxic by-product, homogentisic acid, accumulates in the blood and is excreted in the urine, which turns dark upon air oxidation. Build-up of homogentisic acid also causes cartilage damage (leading to osteoarthritis), coronary heart disease, and kidney stones. Phenylketonuria is caused by a deficiency of phenylalanine-4-hydroxylase, meaning that phenylalanine cannot be converted to tyrosine. The phenylalanine accumulates and is converted instead to phenylpyruvate, which can be detected in the urine. Severe mental retardation occurs if the defect is not diagnosed soon after birth.

> **KEY POINT**
>
> The 20 common amino acids are degraded by different pathways to converge at seven products. The seven products are metabolic intermediates of either glycolysis or the citric acid cycle.

FIGURE 3.14 When amino acids are broken down, the carbon skeletons feed in to different parts of metabolism. Those in pink are the glucogenic amino acids and those in yellow are the ketogenic amino acids

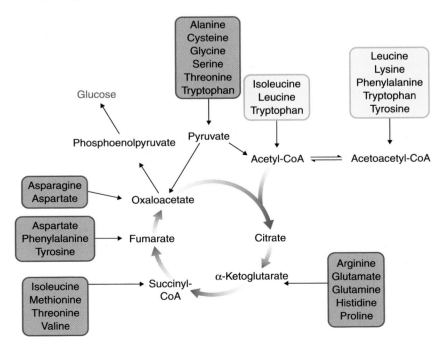

➤ Proteins are grouped into three major groups called the fibrous, globular, and membrane proteins, each of which has different functions which result from their differing structures.

➤ Protein structure can be described in terms of four levels. The primary structure is the sequence of amino acids. The secondary structure explains how a polypeptide chain is further stabilized into an α helix, β sheet, β turn, or Ω loop. The tertiary structure describes how the protein forms a more compact shape that further enhances stability, and the quaternary structure describes how some proteins have multiple interacting subunits that come together to give function.

➤ Enzymes are proteins that have catalytic function by virtue of their active site. They are specific and have affinity for their substrates. Allosteric enzymes contain multiple active sites and multiple subunits. Enzymes stabilize the formation of a transition state and alter the reaction rate but not the equilibrium.

➤ Enzyme inhibition can be irreversible, via modification of the structure of the active site, or reversible, via noncovalent interactions that are easily removed.

➤ Enzymes are vital for metabolism. Metabolic pathways can be anabolic (requiring energy) or catabolic (forming energy). There are three fuels for metabolism: carbohydrates, fats, and proteins.

➤ Glycolysis is the breakdown of glucose to pyruvate. There are a number of enzymes involved and the key stages are regulated to ensure appropriate use of energy resources.

➤ The citric acid cycle is an aerobic process which converts pyruvate into acetyl-CoA and energy (in the form of ATP). It occurs in a cyclical series of reactions involving different enzymes at each stage.

➤ Physiologically, the regulation of glycolysis and the citric acid cycle depends on the tissue, the level of physical activity, and the intake of fuels in the diet. The liver and muscle are the key tissues involved. Depending on the blood glucose level, the body either metabolizes what has been taken in or demands fuels from the stores. However, if too many fuel molecules are eaten, the components are sent for storage.

➤ Lipid catabolism produces fatty acids which undergo β-oxidation. This forms acetyl-CoA, which enters the citric acid cycle.

➤ Amino acid catabolism starts with the removal of the amino group and leaves a carbon skeleton that enters the metabolic pathway at either glycolysis or the citric acid cycle, depending on the amino acid.

FURTHER READING

Ahmed, N. *Clinical Biochemistry*. Oxford University Press, 2010.

This book links theoretical biochemistry to clinical applications, with excellent case studies throughout.

Bender, D.A. *Introduction to Nutrition and Metabolism*. 4th edn. CRC Press, 2008.

This book discusses enzymes and metabolic pathways in great depth, how nutrients are absorbed, and some of the problems that occur when nutrition is incorrect. It also talks about the nutrients not covered in this chapter.

Buxbaum, E. *Fundamentals of Protein Structure and Function*. Springer, 2007.

This book covers all the basics of protein structure and function in great detail.

Chasman, D. *Protein Structure: Determination, Analysis and Applications for Drug Discovery*. Marcel Dekker, 2003.

This book describes how knowledge of protein structure is used by the pharmaceutical industry to design drugs.

Elliott, W.H. and Elliott, D.C. *Biochemistry and Molecular Biology*. 4th edn. Oxford University Press, 2009.

This book has short, focused chapters that cover all the basic principles of biochemistry.

An introduction to drug action

RACHEL AIRLEY

Pharmacology is all about what drugs do inside the body. As pharmacists, that body belongs to our patients, so we have to gain an appreciation of what drugs do in the body (pharmacodynamics) and, of course, what the body does to drugs (pharmacokinetics), so that we may consider the overall effect on a patient. In this chapter we will learn more about where drugs act in cells and how cellular components can become **drug targets** or receptors for drug action. We will also discuss how drugs exert their pharmacodynamic response, with more in-depth coverage of receptor–ligand binding and downstream cellular responses. We will consider specific examples of different receptor types and where they fit in the treatment of diseases, looking at how the expression, activity, structure, and function of these drug targets are important in the causes of diseases and for the medicines we use to treat them.

Learning objectives

Having read this chapter you are expected to be able to:

➤ Understand the drug–receptor interaction, the different types of receptor involved, and how they work.

➤ Know how receptor binding by drugs leads to a downstream clinical response and understand the role of biological signalling pathways.

➤ Appreciate how receptors in different **subcellular** locations and within different tissues and organs may work together.

➤ Understand the difference between receptors and ion channels.

➤ Describe the concept of tachyphylaxis (tolerance) and give examples of drugs showing this effect.

➤ Explain the pharmacodynamic basis of drug addiction.

4.1 Receptor theory

When we talk about a drug receptor we tend to think of a cell with a branched structure sticking out of it where a drug with a complementary structure may bind—something like Figure 4.1. Although this goes some way to representing cell membrane-bound receptors, this is not the whole story. Not only are there many types of receptor, or *receptor families*, but there are also some receptors found in other parts of the cell.

FIGURE 4.1 The basic concept of the drug–receptor interaction

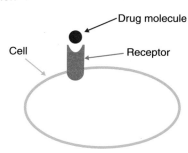

The binding of a drug to a receptor is by a type of chemical bond. The strongest chemical bonds are **covalent**, whereas **electrostatic bonds** vary in strength—ionic bonds being much stronger than hydrogen bonds and van der Waals forces, which are to do with dipole interactions or the way positive and negative charge is distributed between interacting molecules.

 Chapter 2 'Organic structure and bonding', in the **Pharmaceutical Chemistry** book within this series covers the different types of chemical bonds.

Another type of weak force that might come into play in a drug–receptor interaction is the **hydrophobic bond**. These are important as they come about when **lipophilic** drugs interact with the lipid bilayer of cell membranes, so we think about hydrophobic bonds in relation to how drugs interact with membrane-bound receptors or in situations where drugs are taken up (**endocytosis**) or leave (**exocytosis**) a cell.

One aspect of Figure 4.1 is certainly true, in that if a drug binds to a receptor there is a certain amount of homology, or structural similarity, between the shape of the drug and its receptor. We see a similar concept with the 'lock and key' hypothesis when we consider the binding of enzymes and substrates. This feature of drug–receptor binding is very useful, as it confers *specificity* to the interaction. This means that only drug molecules with the correct structure can bind and initiate a response.

 For a recap of the 'lock and key' hypothesis see Box 3.1 in Chapter 3.

> **KEY POINT**
>
> We can translate what we know about the drug–receptor interaction into clinical practice if we can work out which receptors relate to a disease and characterize their shape, so that we can design drugs that specifically alter the response at that receptor.

Unfortunately, the degree of specificity is never quite perfect; one reason for adverse drug reactions may be that the drug is also able to bind to receptors related to quite different biological pathways and physiological functions. We should therefore also consider *selectivity*. Selectivity is achieved when the chemical structure of a drug is such that it fits a certain receptor type but not others. This makes sure that drug–receptor binding leads to the desired clinical effects. Without selectivity the drug may bind to other receptors, leading to unintended consequences such as side effects.

> **KEY POINT**
>
> Although a particular drug might bind to the intended receptor, this receptor could also be found in other tissues or organs in the body and the response the drug induces via these receptors may potentially be harmful or cause toxicity.

4.2 Basic concepts of pharmacodynamics

Having discussed the basics of the drug–receptor response, and now that we know that it is not all about membrane-bound receptors, we will take a look at the different classes of receptor and how they produce a drug response.

Where do we find receptors?

Receptors are proteins that vary in structure, function, and cellular location. Although their function is linked to a response, the mechanism which couples each type of receptor to that response varies. There are broadly speaking five types of receptor, which are illustrated in Figure 4.2. These include the G-protein-coupled receptors; ligand-gated ion channels; voltage-gated ion channels; tyrosine kinase receptors; and the nuclear receptors.

G-protein-coupled receptors

G-protein-coupled receptors are able to interact with extracellular ligands, which may be natural (endogenous) or a drug molecule that simulates some aspect of the endogenous ligand's structure. We find these receptors on the cell membrane, where the ligand most often binds to the exofacial (faces outwards from the cell) domains of the receptor. Ligand activation on the outside of the cell activates G-proteins found on endofacial (inward facing) domains of the receptor, which face the cytoplasm. G-proteins stimulate the activity of enzymes that increase the level of an intracellular second messenger. Second messengers include cyclic nucleotides, such as 3′,5′-cyclic adenosine monophosphate (cAMP) and 3′,5′-cyclic guanosine monophosphate (cGMP), and the phosphoinositides, which are intermediates in the signal transduction pathways regulating many physiological adaptations and functions. Examples of G-protein-coupled receptors include α- and β-adrenoceptors, which bind the neurotransmitter noradrenaline; muscarinic acetylcholine receptors, which bind acetylcholine; and serotonin (5-hydroxytryptamine; 5-HT) receptors.

FIGURE 4.2 **The five different types of receptor**

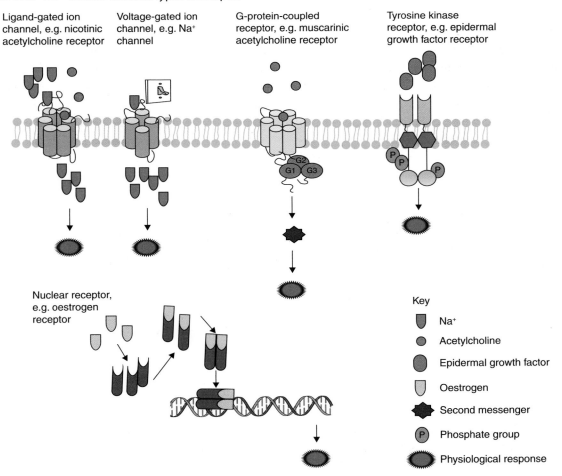

You will read more about these receptors in Chapters 5, 6, and 7.

Ligand-gated ion channels

These are carrier protein complexes made up of several membrane-spanning protein domains. When the correct ligand binds, the protein undergoes a conformational change and a pore or channel is formed through which ions can enter the cell. These include the nicotinic acetylcholine and the γ-aminobutyric acid (GABA) receptors. They are particularly suited to their job as synaptic receptors because they open very rapidly in response to ligand binding, essential for propagating nerve transmission.

 Chapters 5 and 6 cover the role of ligand-gated ion channels in synaptic transmission.

Voltage-gated ion channels

Like ligand-gated channels, these are membrane-spanning protein complexes. Their mechanism for opening is different, though, as they are activated by a change in the electrical potential difference at the cell membrane. We find these in particular at the neuromuscular junction—for example, sodium (Na^+) and potassium (K^+) ion channels; or in muscle cells, where the force of contraction is determined by calcium ion (Ca^{2+}) influx through voltage-gated Ca^{2+} channels. In muscle, at resting membrane potential, the ion

channel will normally be closed and depolarization of the cell membrane initiates opening of the channel. Inhibitors of voltage-gated ion channels include calcium antagonists used to treat hypertension, such as amlodipine, and the gated K^+ channel blocker amiodarone, an antiarrhythmic drug.

 You can read more about membrane potentials in Chapter 5.

Tyrosine kinase receptors

These receptors trigger a host of signal transduction pathways of particular interest to those studying cancer and proliferative disease, as they often trigger the pathways regulating the process of cell division. The structure of the tyrosine kinase (TK) receptor consists of a range of polypeptide components, each with a respective job to do. Common to all TK receptors is an exofacial domain that binds the appropriate ligand, which causes dimerization of the receptor. The external domains of the receptor are connected by a lipophilic bridge through the cell membrane to an endofacial TK enzyme domain. When the receptors dimerize, the two TK domains come together, and are now able to accept phosphate groups. This autophosphorylation step is a typical feature of TK receptors, which results in the activation of the TK enzyme. The function of the enzyme is to then phosphorylate tyrosine residues on downstream signalling proteins. Some examples of these include insulin receptors, the phosphorylation of which results in an increased uptake of glucose, and epidermal

---

BOX 4.1

Cytokine receptors

Cytokines, which are instrumental in immune and inflammatory responses, are involved in many disease states. Cytokine receptors bind cytokines and act in a similar way to TK receptors, except that rather than being activated by an autophosphorylation mechanism they use a separate 'companion' TK called JAK (Janus kinase) to trigger phosphorylation of the signal transduction protein STAT (Signal Transducer and Activator of Transcription). STAT is a transcription factor that can regulate many different types of gene. There are two major

types of cytokine receptor: type 1, for example the interleukin receptors, and type 2, for example the interferon receptors. The genes activated by binding to the cytokine receptors are associated with increased cell division, manufacture of red blood cells, and immunity to viruses. Cytokines themselves may be used as drugs for the treatment of cancer or multiple sclerosis, but drugs that block the cytokine receptor are also undergoing investigation to treat inflammatory diseases such as Crohn's disease (an ulcerative inflammatory bowel disease).

growth factor receptors, which, once phosphorylated, activate an important signalling molecule in the pathology of cancer. Box 4.1 explains a receptor type similar to TK receptors.

In the last decade we have seen many new TK inhibitors approved for clinical use, for example, erlotinib, used to treat lung cancers, and sunitinib, which is used to treat kidney, bowel, and pancreatic cancers.

SELF CHECK 4.1

Give an example of a disease in which tyrosine kinase receptors are important, and describe how they are activated.

Nuclear receptors

The nuclear receptors are quite different from the others as they are found inside the cell, where they usually reside in the cytoplasm. These typically respond to binding of steroid hormones such as oestrogen and androgens (for example testosterone), mineralocorticoids, corticosteroids, and thyroid hormone. Box 4.2 explains the use of corticosteroids in treating various allergies such as hay fever. Ligand binding at the nuclear receptor, just as with the TK receptor, causes them to dimerize or pair up. That is where the similarity ends, however, as the resulting conformational change forms a transcription factor which directly binds to DNA sequences that have regulatory roles or are response elements within gene promoters. These activate or suppress gene transcription (switch genes on or off) in response to certain physiological stimuli such as drugs, hormones, or changes in biochemical metabolites. These receptors are the target of drugs such as tamoxifen, an oestrogen receptor antagonist used to treat some types of breast cancer, and cyproterone, an androgen receptor antagonist used to treat prostate cancer.

 Section 2.3 within Chapter 2 explains the processes of gene transcription.

Drug targets beyond the receptor: up- and downstream effects

The clinical response experienced by a patient once they have taken a medication is determined mostly by what takes place beyond the receptor. Granted, to a certain extent, parameters such as *receptor affinity*, which tells us something about how tightly a drug binds to a receptor, may ultimately affect the dose we use. When we are deciding what the clinical effect of the drug will be, however, we cannot consider affinity without also considering *efficacy*, which describes and measures the downstream response induced by ligand binding to the receptor.

SELF CHECK 4.2

What role do receptors play in the biological response to a drug?

Take a look at Figure 4.3, which represents a cell and the communication pathways between receptor and response. We see the different types of ion channel—on the left a channel that might be a ligand- or voltage-gated

BOX 4.2

Corticosteroid treatments

Nuclear receptors help to explain why corticosteroids are used as 'preventers' in asthma and why corticosteroid nasal sprays need regular use to relieve allergies, and take a few days to work.

When we tell patients that their beclometasone inhaler treats the cause and not the symptoms of hay fever, for example, this is absolutely true. The pharmacological basis of this is that decongestants tend to be sympathomimetics, such as the α-adrenoceptor **agonist** xylometazoline, which work by causing **vasoconstriction** of the leaky blood vessels in inflamed tissue in the nose, thus treating the symptoms. The corticosteroids, however, influence the production of the physiological mediators that actually cause inflammation, such as the **interleukins** and various **growth factors**, and therefore address the cause of the allergy.

FIGURE 4.3 Subcellular location and downstream effects of different types of receptors

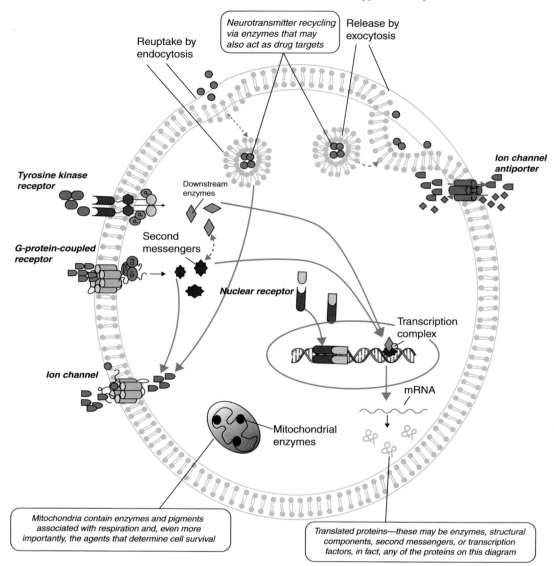

Reuptake by endocytosis

Neurotransmitter recycling via enzymes that may also act as drug targets

Release by exocytosis

Ion channel antiporter

Tyrosine kinase receptor

Downstream enzymes

Second messengers

G-protein-coupled receptor

Nuclear receptor

Transcription complex

Ion channel

mRNA

Mitochondrial enzymes

Mitochondria contain enzymes and pigments associated with respiration and, even more importantly, the agents that determine cell survival

Translated proteins—these may be enzymes, structural components, second messengers, or transcription factors, in fact, any of the proteins on this diagram

carrier of Ca^{2+}, Na^+, or K^+ ions; and on the right, a Na^+/K^+ **antiporter**, which carries two types of ion in opposite directions, depending upon the concentration gradient (we will discuss this in 'Cross-talk between receptor types', later in this section, when we talk about the target of the drug digoxin). We also see two cell membrane-bound receptors on the left, which might be of the tyrosine kinase or G-protein-coupled type, both of which initiate a downstream response via downstream enzyme or second messenger activation, respectively. Finally, we see the nuclear receptor, which is translocated from the cytoplasm to the nucleus, where it acts as a transcription

factor, initiating or repressing transcription of relevant genes by binding to DNA sequences that make up their promoter switch.

What is not immediately obvious, however, is the nature of the response. For example, we might want to know what actually controls the amount of second messenger needed to link binding of the correct ligand to whatever physiological response it triggers; or which genes are switched on when the translocated oestrogen receptor binds to DNA sequences within the promoter. It would also be helpful to understand how the expression of the enzyme intermediates that relay TK receptor phosphorylation to a fully translated response are controlled, and what determines how long receptor signalling is going to last.

For instance, the intermediate enzymes in the TK receptor pathway are also drug targets. The anticancer drug dasatinib inhibits the non-receptor tyrosine kinase SRC, an enzyme intermediate in biological pathways controlling cell division and the transcription of yet more drug targets. In fact, the downstream effects of receptor stimulation upon gene transcription are particularly relevant for the nuclear receptors, as we consider which particular genes they switch on or off and whether or not the proteins they code for may also act as drug targets. Of course enzymes themselves, by virtue of their three-dimensional interactions with their substrate and their ability for allosteric interaction, exhibit kinetics similar to those of ligand–receptor binding, which is why we may count an enzyme as a type of receptor. Mitochondrial enzymes are included in Figure 4.3 as they play a part in cell survival mechanisms and energy production, so are important targets for the design of novel drugs.

The duration of action of the response stimulated by receptor binding also determines the clinical effect. To illustrate this we can consider the receptors at a nerve synapse, where the recycling of neurotransmitters involves their presynaptic uptake and breakdown. If this does not take place there will be high concentrations of neurotransmitter in the synapse, leading to repeated postsynaptic receptor binding and propagation of the nerve signal. In Figure 4.3 we can see that uptake of neurotransmitter (for example acetylcholine) takes place via endocytosis, whereby endosomal enzymes (for example acetylcholinesterase) break down and recycle it. Of course, expression of endosomal enzymes will be a factor in this process, and these enzymes may become drug targets. An example of this strategy is in the treatment of myasthenia gra-

vis, an autoimmune disease in which antibodies block postsynaptic cholinergic receptors, leading to muscle weakness. We treat this disease symptomatically with cholinesterase inhibitors, which increase the level of acetylcholine in the synapse.

 For a more in-depth account of neurotransmitter uptake and breakdown mechanisms see Section 5.3 'How do nerve cells communicate with each other?', in Chapter 5.

Cross-talk between receptor types

Of course the receptors themselves are proteins, which are coded by genes. We therefore have to consider which genes and which receptor-mediated processes themselves trigger the production of the five types of receptor we talked about under 'Where do we find receptors?' in this section. We must also consider that there may be cross-talk between the different types of receptor, where the activation of a response at one receptor leads to changes in transmission of a response signal at another. Figures 4.4 and 4.5 show this in action.

Figure 4.4 illustrates the mode of action of the drug nicorandil, which opens the ATP-dependent K^+ channels on stenotic (narrow) coronary arteries in angina patients, allowing vascular smooth muscle to relax, and increasing perfusion of the heart. Nicorandil causes the K^+ channel to open, letting K^+ ions flood out of the cell. This leaves the cell membrane hyperpolarized (negatively charged). This hyperpolarization of the membrane has downstream effects on nearby voltage-gated Ca^{2+} channels, causing them to remain closed. This means that Ca^{2+} ions will not be able to flow into vascular smooth muscle cells to initiate contraction, relaxing the vascular smooth muscle. This, along with its nitrate-like effect, gives nicorandil its vasodilatory action.

The cardiac glycoside digoxin is an inotropic drug that acts at one ion channel in order to mediate its clinical effect at another. It does this by manipulating the concentration gradient of Na^+ ions inside relative to outside the cell. Ca^{2+} ions enter muscle cells by two mechanisms: an ATP-dependent Ca^{2+} pump and a Na^+/Ca^{2+} exchange pump. We can see in Figure 4.5

FIGURE 4.4 Mode of action of the antiangina drug nicorandil

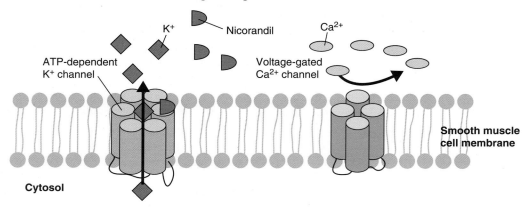

how digoxin increases the force of contraction of the heart muscle by disrupting the concentration gradients needed to direct the activity of the Na^+/Ca^{2+} pump. It does this by binding to the cardiac glycoside receptor on the Na^+/K^+ exchange pump, which normally pumps Na^+ ions out of the cell in exchange for K^+ ions. When digoxin binds to its receptor, however, the pump stops working, so that Na^+ builds up inside the cell cytoplasm. This affects the Na^+/Ca^{2+} exchange pump on another part of the cell membrane. This pump draws Na^+ ions into the cell whilst forcing out Ca^{2+} ions. The Na^+/Ca^{2+} exchange pump will not function effectively unless there is a concentration gradient that favours this movement of ions, that is, a higher concentration of Na^+ ions outside than inside the cell. Of course, that gradient was lost when digoxin stopped the Na^+/K^+ exchange pump from doing its job. Therefore, there is so much Na^+ inside the cell it is impossible for the Na^+/Ca^{2+} pump to do its regular job and pump Ca^{2+} ions out.

FIGURE 4.5 Mode of action of the cardiac glycoside digoxin

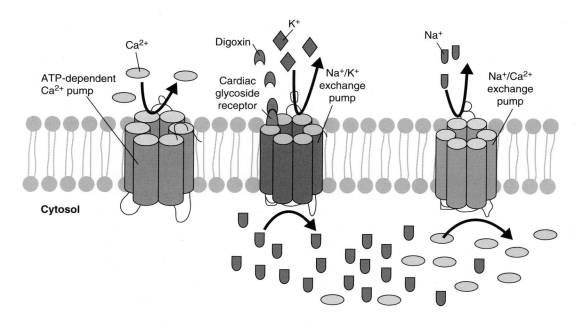

This results in an increased level of Ca^{2+} ions inside the cell, which increases cardiac contractility or the force of contraction of the heart muscle (inotropic effect).

SELF CHECK 4.3

Give an example of how drugs may produce their effects by interacting with ion channels.

Agonism, antagonism and their role in therapeutic responses

Being pharmacists, we tend to think of receptors as being there to make drugs work. Of course this is not true. Receptors are usually bound by endogenous ligands such as hormones, ions, metabolites, or growth factors, where receptors are the cellular link between these ligands and a biological response, whether this is to initiate a homeostatic pathway or perhaps one that initiates growth or development. As we know, pathology, or the study of disease, is about how normal physiology goes wrong to cause disease. Disease usually involves the deregulation of a biological pathway, which will at some point be receptor-driven. In general, one of these biological pathways can either proceed too much or too little, which can have either direct or knock-on effects that cause disease.

KEY POINT

We can interfere with receptor-driven processes by using drugs that act at these receptors to bring that biological pathway back to normal. So, we want a drug that either decreases the action of a pathway or increases it.

A receptor may be bound by an *agonist* or *antagonist*. An agonist is likely to be a structural analogue of the natural or endogenous ligand and will activate the receptor response. We describe an agonist as having *affinity* and *efficacy*. An antagonist is likely to be a molecule that can bind to the receptor and prevent the agonist binding or prevent the subsequent biological pathway from occurring. Antagonists therefore have *affinity* but *no efficacy*.

KEY POINT

Affinity is the ability of the drug molecule to bind the receptor owing to its three-dimensional structure, and efficacy as the relationship between receptor occupancy and response, where the maximum response (known as E_{max}) is induced when the receptors are at maximum occupancy.

Agonism

To compare different agonists, *dose–response curves* are often plotted. These show the relationship between ligand (endogenous or drug molecule) concentration and response. If we consider the dose–response curves shown in Figure 4.6, we can see that drug A has roughly half the maximal response (E_{max}) of drug B. This may be due to differences in the nature of receptor binding or an indication that different receptors are involved. Conversely, a look at drugs B, C, and D tells us that these drugs have a similar E_{max}, which might indicate that similar receptor systems are involved in their mode of action and therefore in the therapeutic response they may elicit in the patient.

Potency is an indication of the concentration of drug needed to produce that maximal response. We usually quantify this concept by calculating the concentration of drug required to achieve half the maximal response. Potency is therefore defined as the effective concentration at 50% of E_{max} (EC_{50}). The higher the concentration required to achieve half of E_{max} the lower the potency of that drug. If we take another look at Figure 4.6, and

FIGURE 4.6 Relationship between efficacy, potency and therapeutic response of four drugs coded A, B, C, and D. Note the *x*-axis is in a logarithmic scale

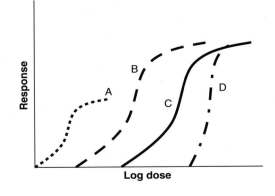

this time consider potency, drug B has a lower potency than drug A, even though it has a higher efficacy. Curiously though, drug C also has lower potency than drug B, despite their maximal effects being similar. Drug D has an even lower potency, yet the maximal response stays the same.

Such dose–response curves allow us to make predictions about the clinical response we might see in patients. Considering pharmacodynamic effects in isolation (in reality this would rarely happen as there are other parameters, for example **pharmacokinetics**, that will influence the overall clinical effect), drugs B and C will be similar in terms of the maximal clinical response they may achieve in a patient. The use of either drug will in fact produce a greater response than drug A. The doses of B and C needed to achieve this response, however, will be different, as a higher dose of C is needed to produce an equivalent clinical response. In a clinical setting this is of little significance; it really only becomes a consideration when the drug is formulated, as the effective dose will determine the strength of the preparation and the **dosage regimen**.

The use of drug D has an additional parameter to consider: a close look at the *shape* of the dose–response curve, which has the steepest gradient, tells us that there is a very narrow dose range between no response at all and the maximal response. Clinically, this means we have to be careful with this drug because it has a very narrow *therapeutic window*, meaning there is little room for manoeuvre between a dose that produces a good therapeutic response and

one that produces toxicity. In such cases, only slight variations in confounding factors existing in patients might lead to differences in the way they respond to drugs. Genetic differences in drug metabolism (**pharmacogenomics**) or even differences in the expression of receptors mediating side effects can tip the balance from safe and effective to toxic drug concentrations.

 Section 1.5 'The effect our bodies have on administered drugs', in Chapter 1, introduces the metabolism of drugs (pharmacokinetics).

KEY POINT

The shape and position of the dose–response curve is determined by affinity, efficacy, and potency and these affect the therapeutic range of the drug.

The effects that binding of an agonist will cause may also be explained by the nature of the agonism involved—a *full agonist* will produce a maximal response whereas a *partial agonist*, although it might well bind to the receptor with a high degree of affinity, has only *partial efficacy* at a receptor when it is fully occupied. The clinical effects of partial agonists vary according to whether they are acting in the presence of other drugs binding to that receptor. If there is a high concentration of a full agonist present, the partial agonist will in fact act as an antagonist. This is because it will be competing for the same receptor as the full agonist, and as such will prevent a maximal response. Meanwhile, in the presence of an antagonist or extremely low concentrations of the

---

BOX 4.3

Tamoxifen: the whole picture

We see a good example of partial agonism with tamoxifen, a partial agonist of the oestrogen receptor. Tamoxifen is used as maintenance therapy in the treatment of oestrogen-dependent breast cancer. In this type of cancer, binding of oestrogen to the oestrogen receptor promotes the growth and multiplication of the tumour cells. Blocking the effects of oestrogen with tamoxifen is therefore an effective long-term therapy to help prevent the return of the disease. In this case, we are taking advantage of tamoxifen's action as an antagonist to prevent oestrogen-linked cell pathways that favour the recurrence of cancer.

But tamoxifen also acts as a partial agonist at oestrogen receptors in the **endometrium** and in bone, and this lack of tissue selectivity means we have to consider the potential effects of tamoxifen at other sites. We therefore have to consider the risk of developing hormone-dependent endometrial cancer via increased activity at these receptors, which is a negative point for tamoxifen. Its effect on bone receptors, however, is beneficial, as the partial agonist effect at the oestrogen receptor in bone maintains bone density, meaning that tamoxifen offers protection against osteoporosis in postmenopausal women.

full agonist, a partial agonist will behave as an agonist, though never achieving the maximal response possible from that receptor. An example of a clinically relevant partial agonist is given in Box 4.3.

Antagonism

As we have mentioned, receptor antagonists are very different to agonists, as they have binding affinity but no efficacy. Therefore, even if there is total saturation of drug binding at the receptor, there is an absence of any downstream biological response. Be careful not to confuse this with an absence of clinical or therapeutic response, as the use of receptor antagonists is appropriate where a disease physiology comes about owing to over-activity of a particular pathway or an unwanted change in the pathway that gives rise to aberrant physiology. In this case, an effective agent will be one that reduces the signs and symptoms of that disease physiology.

There are many types of antagonist, and although they all work differently they all have the same primary purpose of stopping the endogenous agonist activating a response at that receptor. In this chapter we will discuss the three main types of antagonist: *reversible competitive*, *irreversible competitive*, and *non-competitive*. The different effects that antagonist types cause can be explained by considering whereabouts on the receptor the antagonist binds. We can distinguish between them by looking at their inhibitory effect in the presence of agonist at the same receptor, as shown in Figure 4.7.

- *Reversible competitive antagonists* act by binding to the receptor at the same point on the receptor as the endogenous ligand. This reaction is reversible; if the concentration of agonist is high enough it will knock the antagonist out of the receptor and re-establish the receptor response. We can see this effect in Figure 4.7, where for a reversible competitive antagonist, the higher the concentration of agonist also present, the lower the percentage inhibition it produces.

- *Irreversible competitive antagonists* also bind at the endogenous ligand-binding site but the nature of the binding is different. In this case the ligand–receptor complex is covalently bonded and is there-

FIGURE 4.7 The inhibitory action of antagonists in the presence of varying concentrations of agonist helps characterize the receptor–antagonist interaction

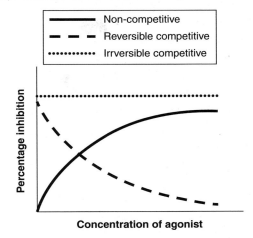

fore irreversible. This means that no matter how high the concentration of agonist, it will not be able to overcome the tight receptor–antagonist interaction. In this case, there is no change in percentage inhibition with agonist concentration, as shown in Figure 4.7.

- *Non-competitive antagonists* bind at an alternative or allosteric site away from the endogenous ligand-binding site. The antagonist will only work, however, if the receptor has already been stimulated by binding of the endogenous ligand. If we take another look at Figure 4.7, we can see that for an non-competitive antagonist, the percentage inhibition rises with the concentration of agonist.

We can represent the effect of reversible and irreversible competitive antagonists slightly differently. Figure 4.8 compares the dose–response curves of an agonist in the presence of these antagonists. These curves show us how the potency and efficacy of an agonist are altered in the presence of fixed concentrations of antagonist. As we discussed earlier, we use EC_{50} to describe potency and E_{max} to represent maximal agonist response. When we plot a dose–response curve in the presence of a fixed concentration of antagonist we can work out which type of antagonist we have by reflecting on the changes in these parameters. The inhibitory effects of a reversible competitive antagonist will be overcome if the concentration of agonist

FIGURE 4.8 Dose–response curves showing the response caused by an agonist alone and the inhibition of the agonist response by reversible and irreversible competitive antagonists

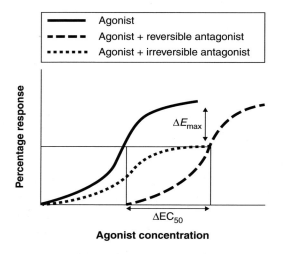

Agonist concentration

is large enough, therefore we see a characteristic right shift in the dose–response curve, resulting in an *increased* EC_{50}, but *unchanged* E_{max}. In the presence of an irreversible competitive antagonist, however, increasing the concentration of agonist will not overcome its inhibitory effect—in fact, the formation of irreversible covalent bonds means that there are now far fewer receptor sites available for the agonist to bind. Therefore we see a *decreased* E_{max} but an *unchanged* EC_{50}. The clinical application of competitive antagonists is presented in Case Study 4.1.

SELF CHECK 4.4

Define the terms efficacy, potency, and E_{max} and describe how these parameters may translate into clinical response.

CASE STUDY 4.1

George suffers from travel sickness and asked the pharmacist for a suitable remedy. She recommended JoyRides® and told him to be careful as it would make him drowsy.

The following day, George came back to the pharmacy concerned about his travel sickness tablets. He had read the box and noticed that they contain a drug called hyoscine which he said his girlfriend takes for irritable bowel syndrome. He was worried that the drugs are 'too strong for just travel sickness'. The pharmacist explained to George that although the active ingredient of the two formulations are similar, they are formulated to reach different parts of the body. The pharmacist then went on to reassure George that the JoyRides® tablets were appropriate for management of travel sickness and that he should take them as instructed.

REFLECTION QUESTIONS

1. Explain how hyoscine is used to manage travel sickness as well as treat irritable bowel syndrome (IBS).

2. Why couldn't George take his girlfriend's IBS tablets to treat his travel sickness?

3. Explain how hyoscine works as an antimuscarinic agent.

Answers

1 Hyoscine is a reversible competitive antagonist of the muscarinic acetylcholine receptors. These receptors are widespread throughout the body. In the JoyRiders® the hyoscine will affect the way signals are transmitted from the vestibular apparatus in the inner ear to the parts of the brain controlling the vomiting reflex. Whereas, hyoscine in the tablets used to treat IBS (Buscopan®), will affect gastric motility and dry up various digestive secretions. So, the two medications work in different ways.

2 The active compound in JoyRiders® is a different hyoscine salt to that in the IBS tablets. The IBS tablets have hyoscine butylbromide whereas JoyRiders® contain hyoscine hydrobromide. The difference is that hyoscine butylbromide is a quaternary ammonium compound and is therefore is less lipophilic than the hydrobromide salt. Therefore it will not cross the blood-brain barrier and cannot work on the vomiting reflex in the brain.

3 Hyoscine is a competitive antagonist of the muscarinic acetylcholine receptors. Reversible competitive antagonists bind to the receptor at the same point as the endogenous molecule. In the case of the muscarinic receptors, this endogenous molecule is acetylcholine. When hyoscine binds to the receptor it stops acetylcholine from binding to the receptor, thus prevents the normal physiological process from occurring.

4.3 Adverse drug reactions, drug interactions, and drug tolerance

We define an *adverse drug reaction* (side effect) as a harmful or unintended response to a drug. A *drug interaction*, however, refers to the effect we might see when two drugs are administered concurrently. As pharmacists, these two aspects of pharmacology are important in helping us to carry out a clinical check, explain any unwanted symptoms or side effects, and prevent failure of the treatment or harm to the patient. Drug tolerance, or tachyphylaxis, is a phenomenon that takes place when a drug response is decreased.

Adverse drug reactions

Adverse drug reactions may be *predictable* or *unpredictable*, where predictable reactions may be explained by our knowledge of the drug pharmacodynamics. For instance, we know that the therapeutic effect of the antihypertensive drug lisinopril comes about via its action as an inhibitor of the angiotensin-converting enzyme (ACE). The function of ACE is to convert angiotensin 1 to angiotensin 2, a potent vasoconstrictor. By inhibiting the production of this vasoconstrictor we reduce vascular resistance, and with it, blood pressure. Unfortunately, this is not the only effect of ACE inhibition, as ACE is also used in the body to break down an endogenous **bronchoconstrictor** called bradykinin. If we prevent ACE from carrying out this function, a patient will have increased levels of bradykinin, causing bronchoconstriction and the persistent dry cough often experienced by patients taking lisinopril.

This side effect is easy to predict as it is directly associated with the therapeutic mode of action—in fact, the therapeutic and adverse effects are initiated by the same drug target. An unpredictable adverse drug reaction, however, results from a pharmacological effect that is unrelated to the therapeutic mode of action. A good example of this is the pulmonary toxicity we see in patients being treated for cancer with bleomycin.

 For more details of the effect of kinins, such as bradykinin, on the body see Section 7.4 'Kinins', in Chapter 7.

Drug interactions

Drug interactions, like adverse drug reactions, may be predictable or unpredictable. Sometimes we do not know why two drugs interact and we only know that they do so from clinical or preclinical data. Drugs may interact via pharmacodynamic or pharmacokinetic mechanisms and the interaction may be in the form of combined toxicity, combined therapeutic effect, or reduced therapeutic effect. A pharmacodynamic interaction may be *additive*, where two drugs act on the same receptor, or *synergistic*, where they act on different receptors or induce their response via separate biological pathways. Pharmacokinetic interactions take place when one drug has an impact upon the pharmacokinetic characteristics of another.

 Pharmacokinetics is introduced in Section 1.5 'The effect our bodies have on administered drugs', in Chapter 1.

Many pharmacokinetic interactions we come across involve a drug-induced alteration in the activity of the hepatic drug metabolizing enzymes cytochrome P450 oxidoreductases. These enzymes come in many isoforms, and certain drugs are metabolized in the presence of certain isoforms. A good example of this effect is the anticoagulant warfarin, which is metabolized by the cytochrome P450 2C9 enzyme. When warfarin is co-administered with drugs such as cimetidine (a histamine H_2-receptor antagonist used to treat peptic ulcers) or amiodarone (an antiarrhythmic agent), which inhibit this enzyme, serum warfarin levels are potentially raised and can lead to excessive bleeding.

FIGURE 4.9 A clockwise hysteresis loop demonstrates drug tolerance (tachyphylaxis)

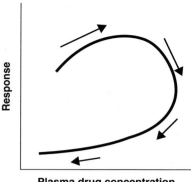

Drug tolerance and variation in response

Drug tolerance (tachyphylaxis) may be due to either pharmacokinetic parameters that interfere with the availability of the drug at the receptor site or changes within the receptors themselves. A *hysteresis loop*, such as Figure 4.9, demonstrates how receptor-mediated tolerance develops. Initially, increasing the drug concentration will give rise to a larger response. In time, however, continued use of the drug causes changes to take place within the receptors, and even their second messengers, which decrease the effect of the drug. This means that gradually, increasingly high doses will have to be used to achieve the same effect. We see tolerance to drugs such as nitrates, used for the treatment of angina, as explained in Box 4.4 and in the misuse of drugs such as diamorphine and cocaine. Tolerance may be particularly dangerous in drug misuse, as

BOX 4.4

Tolerance to nitrates

The basis for tolerance to nitrates used in angina has been studied extensively because of the numbers of patients that experience this effect. Nitrates are used to dilate coronary arteries in patients with angina. The pharmacological basis of nitrate tolerance is complicated, but it has been suggested that a causative factor is the production of **superoxide radicals** as by-products of nitrate activation. In turn, these cause increased sensitivity to endogenous vasoconstrictors, which counteract the vasodilatory effect of nitrates. They also inhibit mitochondrial aldehyde dehydrogenase, which triggers the biological process that converts nitrates to nitric oxide. Tolerance to nitrates usually develops with chronic exposure and is a particular problem in patients using long-acting nitrates or transdermal patches, where its onset can be rapid. In practice, we get around this problem by adjusting the dosage regimen so that serum levels are allowed to drop for a few hours during the day. Patients should remove glyceryl trinitrate patches for a few hours a day, and extended-release oral preparations given twice daily should be given 8 h rather than 12 h apart to allow a nitrate-free period.

increasing doses will be needed to achieve the same feelings of well-being, known as a 'high', increasing the risk and extent of adverse effects such as respiratory depression.

4.4 Drug misuse and addiction

As pharmacists, we are involved in the supply of medicines to drug misusers. An understanding of the pharmacological basis of drug addiction is useful so that we can understand how it develops and how it may be treated. It is also valuable to understand the rationale behind the way drug addiction is treated and the symptoms experienced by the patients involved.

Drug dependence evolves when reward pathways in the brain link the pharmacological activity of certain drugs with the biological pathways within certain areas of the brain associated with mood, causing feelings of euphoria or a 'high'. The drug misuser then becomes dependent on the drug, developing a compulsion to take it to avoid withdrawal symptoms.

KEY POINT

Drug addiction is a result of interference with the function of neurotransmitters involved with reward pathways in the brain. Stimulation of these reward pathways leads to the release of neurotransmitters, which cause feelings of well-being often referred to as a 'high'.

For a more thorough understanding of how particular drugs act on these reward pathways, take a look at Figure 4.10. Two neurotransmitters involved in the reward pathways are dopamine and serotonin (5-HT).

 Serotonin is covered in more detail in Section 7.3 'Serotonin', in Chapter 7.

The drugs featured in Figure 4.10 lead to increased concentrations of these neurotransmitters in the synapse, increasing stimulatory activity. Drugs such as methamphetamine and cocaine may have a direct

FIGURE 4.10 Illustration of how methamphetamine, cocaine, and diamorphine interact with neurotransmitters in the brain involved with feelings of euphoria and well-being

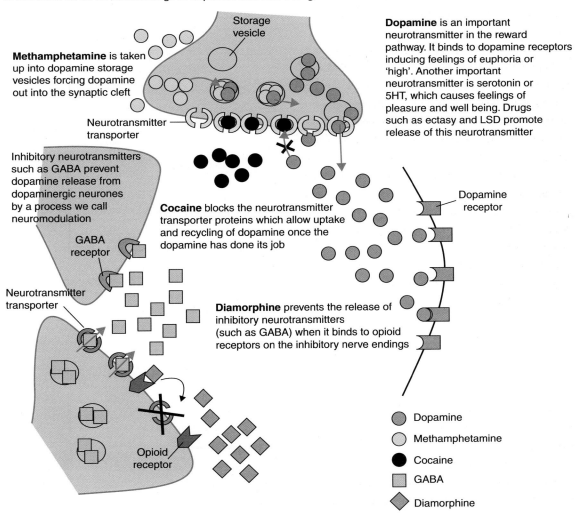

Dopamine is an important neurotransmitter in the reward pathway. It binds to dopamine receptors inducing feelings of euphoria or 'high'. Another important neurotransmitter is serotonin or 5HT, which causes feelings of pleasure and well being. Drugs such as ecstasy and LSD promote release of this neurotransmitter

Methamphetamine is taken up into dopamine storage vesicles forcing dopamine out into the synaptic cleft

Storage vesicle

Neurotransmitter transporter

Inhibitory neurotransmitters such as GABA prevent dopamine release from dopaminergic neurones by a process we call neuromodulation

GABA receptor

Neurotransmitter transporter

Cocaine blocks the neurotransmitter transporter proteins which allow uptake and recycling of dopamine once the dopamine has done its job

Dopamine receptor

Diamorphine prevents the release of inhibitory neurotransmitters (such as GABA) when it binds to opioid receptors on the inhibitory nerve endings

Opioid receptor

- ⬤ Dopamine
- ◯ Methamphetamine
- ● Cocaine
- ▢ GABA
- ◆ Diamorphine

83

effect upon dopamine levels. Figure 4.10 shows how methamphetamine interacts with storage vesicles in the presynaptic nerve terminal to force more dopamine into the synaptic cleft. Cocaine however, blocks the transporter proteins from removing dopamine from the synaptic cleft thus prolonging its effects.

Other drugs, such as diamorphine and the active component of cannabis, tetrahydrocannabinol (THC), bind to their respective receptors on inhibitory neurons and inhibit the release of GABA, an inhibitory neurotransmitter that usually reduces the release of dopamine. This is a good example of neuromodulation.

> **KEY POINT**
>
> Although we now have a much better understanding of the pharmacology of drug addiction, drug dependence is not necessarily a physiological state as it is believed that psychological dependence, where drug misusers repeatedly use drugs for personal satisfaction, may also be an issue.

> **SELF CHECK 4.5**
>
> Describe how people may become addicted to drugs.

4.5 Neuromodulation

The physiology of the peripheral and autonomic nervous systems, the neurotransmitters involved, and how we exploit this system therapeutically will be covered elsewhere in this book. *Neuromodulation* is the regulation of neuronal activity of one neuron by means of a neurotransmitter produced by another. It is possible because the terminal of one neuron can make a synapse with a number of neurons producing different neurotransmitters.

 See Chapters 5 and 6 for further details of the peripheral and autonomic nervous systems.

Looking at Figure 4.11 we can see how a neuron from the sympathetic nervous system, which produces noradrenaline, innervates smooth muscle. Meanwhile, another terminal of this neuron forms a synapse with a parasympathetic acetylcholine-producing neuron stimulating an exocrine gland. In this case, the sympathetic nerve terminal acts to dampen the stimulatory effect of the parasympathetic nerve on this gland. We might see this effect where it is necessary to fine-tune the production of sweat or saliva. Of course, this effect can happen the other way round; the diagram shows how a terminal of a parasympathetic nerve may also fine-tune the effect of sympathetic stimulation on smooth muscle. This might affect, for example, the vasoconstriction of blood vessels in the skin, resulting in the way we look pale when we are nervous.

Neuromodulation means that when we are considering the effect of drugs which are agonists or antagonists of neurotransmitters, we have to consider both the direct effects on the function of the organs they innervate and the effect of boosting or blocking neuromodulation.

FIGURE 4.11 Neuromodulation as a mechanism of fine-tuning nerve transmission. PG = prostaglandin
Source: Reproduced from Rang, H.P., Dale, M.M., Ritter, J.M., Flower, R.J., and Henderson, G. *Rang and Dale's Pharmacology.* 7th edn. Churchill Livingstone, 2012.

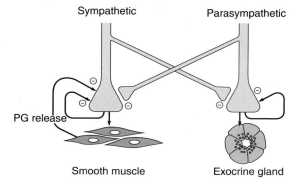

➤ The action and uses of drugs depend upon drug–receptor interactions. The clinical response to these drugs, or their usefulness as medicines, in turn depends upon the way they bind to a range of receptors.

➤ There are five main groups of receptor: G-protein-coupled, tyrosine kinases, nuclear receptors, ligand-gated ion channels, and voltage-gated ion channels.

➤ Different receptors become important drug targets, depending upon their influence on the causes and symptoms of various diseases. This allows us to design drugs which specifically target these diseases. These drugs bind the receptors that are associated with biological pathways fundamental to the alteration in physiological process characteristic of the disease.

➤ Drugs may be receptor agonists, which mimic the effect of natural or endogenous receptor ligands, or receptor antagonists, which block their effect.

➤ Drugs may have unwanted pharmacological and clinical effects, such as adverse drug reactions and interactions between one or more drugs. These processes are unintended but may also be largely receptor-mediated.

➤ Drugs may also lose their effect, a phenomenon called tolerance, which is a particular problem with nitrate agents used in patients with angina.

➤ Drugs such as the opiates, cocaine, and cannabis may be misused, leading to addiction. This is mediated by reward pathways in the brain which involve the neurotransmitters dopamine and serotonin.

FURTHER READING

Camerino, D.C., Tricarico, D., and Desaphy, J.F. Ion channel pharmacology. *Neurotherapeutics* 2007; 4: 184–98.

This article reviews the role of ion channels in cellular processes involved with a range of diseases and how they can be used as targets, particularly where the discovery of different forms of the gene for ion channels (polymorphisms) means we can develop drugs with specific effects.

Levenson, A.S. and Jordan, V.C. Selective oestrogen receptor modulation: molecular pharmacology for the millennium. *Eur J Cancer* 1999; 35: 1628–39.

This review article describes the biology of the oestrogen receptor, its role in a range of diseases, and how it may be targeted by drugs.

Gschwind, A., Fischer, O.M., and Ullrich, A. The discovery of receptor tyrosine kinases: targets for cancer therapy. *Nat Rev Cancer* 2004; 4: 361–70.

The discovery and characterization of the receptor tyrosine kinases and their role in diseases such as cancer and inflammatory disorders have led to the design and discovery of many novel targeted drugs. This review article offers a concise and informative account of how receptor tyrosine kinase inhibitors have changed the way we treat these diseases.

Rang, H.P., Dale, M.M., Ritter, J.M., Flower, R.J., and Henderson, G. *Rang and Dale's Pharmacology*. 7th edn. Churchill Livingstone, 2012.

This pharmacology textbook provides a wide ranging account of the different types of receptors and their associated physiological responses.

Camí, J. and Farré, M.D. Drug addiction. *N Engl J Med* 2003; 349: 975–86.

This is a useful article that reviews the mechanism of drug addiction and offers a comprehensive account of the receptor-mediated pathways and parts of the brain involved.

5

Communication systems in the body: neural

NEIL HENNEY AND PETER PENSON

Think about any time you have felt pain, or sensed extremes of heat or cold. Think about your ability to move your body about, to wriggle your fingers, to play a musical instrument or a computer game. These tasks that you carry out all of the time, which fundamentally enable you to survive by finding food and moving away from danger, rely on very rapid communication between cells and organs. This job is performed admirably by your nervous system. The brain and spinal cord make up the central nervous system (CNS), which is the control centre of the body, while the peripheral nervous system carries messages from the CNS to the body's cells and organs, and back to the CNS for processing.

This chapter will look at the structure and function of the peripheral nervous system. We will examine how cells and organs can communicate very rapidly with one another, at distances as long as you are tall. We'll take a look at the stimuli that make your skeletal muscles contract, and consider how signals travel from your sense organs to your brain, and how messages of pain are communicated. We will consider what happens when things go wrong with the peripheral nervous system, resulting in severe movement disorders. And finally, we'll look at the various very useful drugs that act on the peripheral nervous system, including the local anaesthetics we could not do without at the dentist's surgery, and muscle relaxants that keep us still during complex surgical operations.

Learning objectives

Having read this chapter you are expected to be able to:

- ➤ Describe the roles of the peripheral nervous system.

- ➤ Explain how signals are conducted through neurons and across synapses, including the neuromuscular junction.

- ➤ Explain how excitation–contraction coupling leads to muscular contraction.

- ➤ Describe the structure of the peripheral nervous system.

- ➤ Explain how drugs can affect signal transmission through nerves and across the neuromuscular junction, and how these effects can be useful in medical treatments or surgery.

5.1 The need for rapid communication

Take a note of yourself reading this chapter. Perhaps you are sitting quietly in your room, in a comfy chair with a mug of hot coffee and a chocolate biscuit nearby. Maybe you have the radio on low in the background. You tap your foot softly to the rhythm of a well-known tune while you make brief notes to help you study. You reach out to grab your mug, but Ouch! it's too hot to hold, and you quickly put the coffee down. This all seems very mundane, everyday stuff. But how is your body able to do all these things? How does your nose signal your brain that coffee is nearby, and how does the skin on your hand signal your brain of the painful heat of the mug? How does the signal to tap your foot or handwrite notes travel to your limbs and order your muscles to contract? The answer lies in the impressively specialized and organized network of electrically active cells that form the nervous system.

In fact, the body has two major means of communicating between cells and organs over long distances: either using circulating chemical messengers (hormones) or by nervous transmission. The means of communication that you will learn about in this chapter is the nervous system, which uses electrical signals to rapidly send discrete messages between specific cells and organs. You might compare the hormonal system to a delivery boy posting advertising leaflets for double glazing through everyone's letterbox in a town. All households receive the leaflet (the message) and each may respond differently by ignoring the message, destroying it, or acting and buying new windows. On the other hand, you might compare the nervous system to a salesman from the glazing firm telephoning individual houses, targeted because they are known and likely to act on the message. Individuals in rented houses or high-rise flats are not targeted as this would not achieve any sales for the firm. The nervous system message is directly targeted and discrete.

 If you read through Chapter 7 you will learn about the other means of long-distance communication using autocoids and hormones.

Although the slower, blanket messages of the hormonal system work perfectly well when rapidity is not required, slow messages are not sufficient to signal a muscle to contract or for a painful message to warn the brain of injury—the messages are simply not quick enough, as demonstrated in Box 5.1.

BOX 5.1

The nervous system: the need for speed

Consider a virtuoso pianist playing a complex, fast piece of music during a concert. The brain needs to send messages very rapidly to the muscles of the arms and fingers to contract and then allow them to relax multiple times per second, in order to control movement and hit the right keys at the right time. Hormonal messages would be far too slow for this and each muscle signal, with feedback, could take minutes. After a minute or two of this, I'm sure you (and certainly I) would have left the concert hall.

KEY POINT

Nerve signals travel over long distances very rapidly, in milliseconds (ms), and are discretely targeted.

Divisions of the nervous system

The nervous system is the body's means of rapidly communicating messages between cells, and it is categorized into two major parts according to its functions. The first is the body's control centre, comprised of the brain and spinal cord, and is known as the *central nervous system* or *CNS*. The second part is the *peripheral nervous system (PNS)*, which carries signals out from the CNS, branching all across the body, and also carries signals back from the body's cells and organs to the CNS for processing.

The PNS can be further divided into two functional parts:

FIGURE 5.1 Hierarchy and divisions of the nervous system. The central nervous system (CNS) and peripheral nervous system (PNS) are in constant communication with each other. The CNS receives sensory signals from the PNS for processing, and sends messages out through the PNS to invoke an action, such as gland secretion or muscle contraction

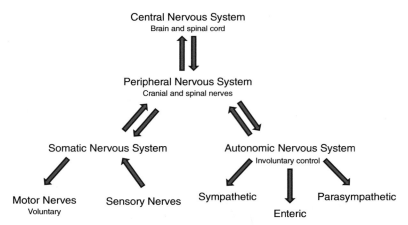

- The *somatic nervous system*, which signals voluntary muscle movements and receives sensory messages from external stimuli.

- The *autonomic nervous system*, which carries signals for involuntary control of bodily functions, such as digestion and heart rate.

Take a look at Figure 5.1, which shows the divisions of the nervous system. In this chapter you will learn about the somatic functions of the PNS.

 You can find out more about the autonomic nervous system in Chapter 6.

SELF CHECK 5.1

Why are nerve signals ideally suited to physiological functions such as muscle movement and coordination?

5.2 An exciting journey: how are signals transmitted?

Cells of the nervous system

Two types of cell make up the PNS. The cells responsible for transmitting signals are called *neurons* (nerve cells), and the cells which support the neurons in their task are called *neuroglia* (glial cells). Let's look at each of these in turn.

Neurons

Neurons and muscle fibres that carry electrical signals are *excitable cells*—they are electrically excitable. Neurons have a specialized structure that enables them to carry electrical signals rapidly across long distances. The cell bodies (called the *soma*) of neurons, containing organelles like the nucleus, are found in the brain and spinal cord and also accumulate in swellings called *ganglia* throughout the body. Look at the diagram of a neuron in Figure 5.2. Branching out in all directions from the soma are fine projections called *dendrites*, which receive incoming signals from other neurons nearby. Extending out from the soma is also a much longer, single projection called an *axon*. The axon is the part of the neuron that carries outgoing signals to other cells. At the beginning of the axon is the *axon hillock*, which is where the

FIGURE 5.2 Diagram of a neuron

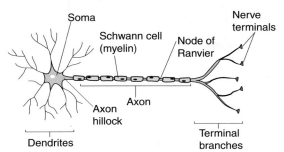

electrical impulses come together before entering the axon. You can see that towards the end of the axon, at the *nerve terminal*, the axon branches out so that a number of cells can receive the same signal together. Each axon terminus is swollen slightly where it meets its target cell. The point of contact between neurons, or between a neuron and the cells of a target organ, is called a *synapse*.

Neuroglia

Neuroglia do not carry electrical signals (they are non-excitable cells), but support the neurons in their task. Many neuronal axons will be covered and insulated by multiple layers of a fatty material called *myelin*, which in the PNS is formed by a type of neuroglia called *Schwann cells*. These cells grow to surround some neuronal axons to support them and also greatly increase the speed of electrical impulses along these myelinated axons. Look at Figure 5.3, which shows you how Schwann cells wrap themselves around neuronal axons, much like cling film is wrapped around its dispensing roll, forming multiple layers of myelin sheath. Long axons may be surrounded by many Schwann cells, with only small gaps

FIGURE 5.3 (A) Diagram of a Schwann cell forming the myelin sheath around an axon, with nodes of Ranvier between neighbouring Schwann cells. (B) A cross section of an axon, showing how Schwann cells wrap around an axon to form the insulating myelin sheath

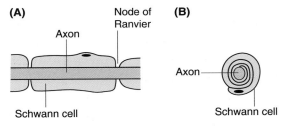

in between called *nodes of Ranvier*. These nodes are the only non-insulated points along the axon in contact with the extracellular fluid, which is important for signal conduction.

> **KEY POINT**
>
> Neurons are excitable cells that carry electrical impulses, whereas neuroglia are supportive and insulating cells that help the neurons in their task.

> **SELF CHECK 5.2**
>
> What is the name of the neuroglial cells that form the myelin sheath around axons in the PNS?

How does a neuron conduct an electrical impulse?

Each signal transmitted along an axon is an electrical wave of positive charge flowing across the axon membrane, which is known as an *action potential*. But before we look at action potentials we should briefly consider the normal state of electrical charge across the membrane—known as the *resting membrane potential*.

Maintaining a resting membrane potential

Every cell in the body maintains different concentrations of ions inside and outside the cell. Ions are charged atoms and cannot cross through the cell membrane of cells on their own. Cells work hard to keep the sodium ion (Na^+) concentration low inside the cells, using energy from **adenosine-5′-triphosphate (ATP)** to pump sodium ions out of the cell against its **electrochemical gradient**. This pump works by exchanging three sodium ions going out for two potassium ions (K^+) coming in, and is therefore called the *sodium–potassium pump*. So as a result, inside the neuron there is a low concentration of sodium ions but a high concentration of potassium ions.

Cell membranes also contain special proteins which form pores, or channels, through the membrane. These *ion channels* allow ions to move into or out of the cell by diffusion down their electrochemical gradient, and are selective in which ions they allow to pass through.

89

So, sodium ion channels allow sodium ions to flow *into* the cell, and potassium ion channels allow potassium ions to flow *out* of the cell. These ion channels can be open or closed, and are therefore said to be *gated*—just like the gate to your garden enables you to stop people or animals moving in or out. If the channels are closed, as they normally are when there is no action potential, ions cannot flow through the channels. The opening and closing of these channels is controlled by changes in the membrane potential, and therefore they are said to be *voltage-gated ion channels*. Other forms of ion channel exist in which the gate is controlled by other stimuli, including those controlled by specific molecules or *ligands* (including drugs) acting on receptors with an integrated ion channel, and these are called *ligand-gated ion channels*.

 You can read more about ligand-gated and voltage-gated ion channels, among other receptor types, in Section 4.2 'Basic concepts of pharmacodynamics', within Chapter 4 of this book.

At rest (no action potentials), a small amount of potassium ions leak out of the cell through *potassium-selective leak channels*. These leak ion channels are not sensitive to voltage changes, and remain open most of the time. This leak of potassium ions out of the cell means that the inside of the membrane becomes *negatively charged* compared with the outside of the membrane, as more positive charge flows out than flows in. It is this outward leak of positive charge that gives rise to the resting membrane potential, which in neurons is approximately –70 mV (millivolts).

> **KEY POINT**
>
> The sodium–potassium pump maintains low levels of sodium ions and high levels of potassium ions in cells. But the continuous leak of positively charged potassium ions out of cells results in a negative resting membrane potential.

Action potentials

In order to generate an action potential a signal must stimulate the cell to do so, and this signal must be strong enough to pass a certain level—the *threshold potential*. If the signal is too small (subthreshold), an

action potential will not be generated. Box 5.2 further explores these stimulating signals. If the signal is larger than the threshold, an action potential will be generated, but even larger signals will only generate the same action potential impulse, however large. Compare this to ringing a doorbell. Press too gently and the bell will not ring. Press the button with more force and the bell sounds in the house. But press harder still and you only get the same response—the bell sound does not change or ring louder. The generation of an action potential is therefore said to be *all-or-nothing*. Either an action potential is generated or it is not, depending on the whether the stimulating impulse reaches the threshold potential.

Action potentials travel along axons only, and not through the soma. During an action potential, the axon membrane potential briefly reverses in polarity (depolarizes) from the resting membrane potential of about –70 mV, rising to about +50 mV at its peak. Look at Figure 5.4, which shows the changes in membrane potential during an action potential. To initiate an action potential, an initial stimulus of small electrical impulses gather at the axon hillock. Here, the combined impulses cause a small number of voltage-gated sodium ion channels in the axon to open, allowing sodium ions to move into the *axoplasm* (the inside of the axon) down their electrochemical gradient. In doing so, the axon membrane becomes further depolarized (more positive, as positively charged sodium

FIGURE 5.4 Graph showing the change in axon membrane potential during an action potential, with associated voltage-gated ion channel open/closed states

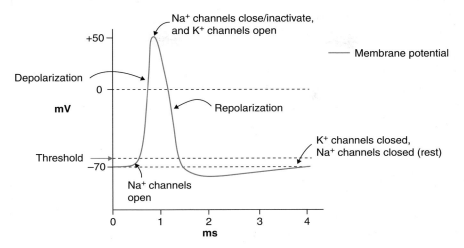

ions have moved into the cell) causing more voltage-gated sodium ion channels to open. This **positive feedback mechanism** continues until all voltage-gated sodium ion channels are open as the axon potential reaches its peak at about +50 mV. At this stage, the membrane is briefly much more permeable to sodium ions than to potassium ions. Some of the initial research showing the importance of ion channels for action potential propagation is considered in Box 5.3.

Voltage-gated sodium ion channels can exist in three activation states:

- *Active*: the channel is open and ions may flow through.
- *Resting closed*: the channel is closed, so ions cannot flow but the channel can be activated.
- *Inactive*: the channel is closed, so ions cannot flow and it cannot be activated.

Voltage-gated sodium ion channels are not stable in their open/active state, and quickly begin to inactivate. This causes a drop in the permeability of the axon membrane to sodium ions—stops them flowing into the axoplasm. At the same time, voltage-gated ion potassium ion channels open in response to membrane **depolarization** and potassium ions begin to move out of the axon down their electrochemical gradient. The movement of positively charged potassium ions out of the cell causes the membrane potential to become more negative again. As the membrane **repolarizes** towards its resting level of about –70 mV, the voltage-gated potassium ion channels close again. Once the membrane is at

resting potential, the inactivated sodium ion channels can return to their resting closed state.

This rapid reversal of the membrane potential is the action potential, and each lasts only about 2 ms, meaning there can be lots of quick bursts of action potentials in a short space of time, to enable ongoing

BOX 5.3

Understanding the action potential: worthy of a Nobel Prize

The remarkable work carried out by scientists during the pre- and post-war years in Britain led to our current understanding of the importance of sodium and potassium ions in action potentials. Using the giant axon of the squid, a number of experiments demonstrated that axon membranes have different permeabilities to different ions, and the importance of the change in sodium ion permeability during the axon potential was recognized in 1949 by Alan Hodgkin and Bernard Katz. In 1952 these two scientists, along with Andrew Huxley, showed that ionic permeability varies with time and voltage through the axon potential, by using electrodes placed into the axon to measure ionic currents (the voltage-clamp technique). Hodgkin and Huxley proposed the idea of ion channels (a novel idea) to allow ions through the axon membrane discretely. Such was its importance to science, this work culminated in the 1963 Nobel Prize in Physiology or Medicine for Hodgkin and Huxley.

sensation or muscle contraction. This is the essence of neuronal communication.

> **KEY POINT**
>
> Action potentials are electrical impulses that travel along axons. They are moving waves during which there is a brief reversal in membrane potential caused by movement of a small amount of sodium ions into the axon.

What is especially important to note is that although a little potassium and sodium ions have exchanged places, the amount of sodium and potassium ion exchange required to significantly alter the membrane potential and cause an action potential is very tiny. The overall concentrations of sodium and potassium ions both inside and outside the axon remain virtually the same, and the gradients are restored by the sodium–potassium pump.

> **SELF CHECK 5.3**
>
> Which ion channel is responsible for propagating action potentials along axons?

Refractory periods

Immediately after each action potential is a short period of time when it is impossible to stimulate another action potential. This is known as the *absolute refractory period*. This happens because the sodium ion channels are inactivated and must spend a brief period at the resting membrane potential before they can revert to their resting closed state. Once some of the sodium ion channels begin to return to their resting state, but before *all* sodium ion channels have done so, an action potential may be initiated but only if a much larger stimulus opens all the available (resting) sodium ion channels at once. This is because some of the potassium ion channels have not yet closed and therefore make the task of depolarizing the membrane towards threshold potential, by the sodium ion channels opening, a little more difficult. This second period is known as the *relative refractory period*—an action potential can be initiated but only by a large stimulus.

How are action potentials propagated along axons?

An action potential travels as a wave along the axon, away from the cell body towards the axon terminals, much like a Mexican wave travels around a crowded stadium. Look at Figure 5.5, which demonstrates this. When an action potential is initiated in the axon hillock, the membrane potential rises from –70 mV to +50 mV in that small, localized region—the *active zone*. As the membrane potential approaches +50 mV, a potential difference is created between the active zone and further along the axon. A tiny current flows between these areas, which begins to depolarize the membrane further along the axon, causing the voltage-gated sodium ion channels to open here, and so the action potential moves along to this new region. This new zone of activity then invades the next region, and so on, as the action potential travels the entire length of the axon.

The action potential can only travel in one direction, away from the cell body, because the previous sodium ion channels inactivate as the action potential moves on, preventing further depolarization. By the time the sodium ion channels have reverted to their resting state, the action potential wave has moved along the

FIGURE 5.5 Diagram showing the reversal of membrane charges during action potential propagation along unmyelinated and myelinated axons. In unmyelinated axons, the action potentials travel as a continuous wave. In myelinated axons the action potential jumps along the axon from one node of Ranvier to the next. This enhances conduction speeds in myelinated axons

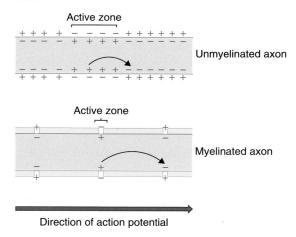

Direction of action potential

axon. This ensures that action potentials only travel in one direction.

> **KEY POINT**
>
> Action potentials only travel away from the cell body of neurons because voltage-gated sodium ion channels quickly inactivate, forcing the wave in one direction.

The speed of action potentials

Generally, action potentials travel faster in larger diameter axons. But the way action potentials travel depends on whether the axon is myelinated. In unmyelinated axons, action potentials are propagated as already described. The entire length of the axon membrane is in electrical contact with the extracellular fluid, so the action potential travels as a smooth wave along the axon. But in myelinated neurons the axon membrane is coated with insulating myelin, except for at the nodes of Ranvier. Look at Figure 5.5, which demonstrates this. At these nodes, electrical contact is made between the axon membrane and the extracellular fluid, but everywhere else the axon is insulated by myelin. This means that a travelling action potential has to jump from node to node, and the great advantage of this is that it allows an action potential to travel much faster. This is known as *saltatory conduction*. In fact, action potentials may travel at only around 0.5 m/s along unmyelinated axons, but in excess of 100 m/s along large myelinated axons—that is about 220 miles/h (360 km/h), the maximum speed of a Formula One car!

A disease such as multiple sclerosis, in which axons slowly lose their myelin coating, results in slower action potential conduction because action potentials in demyelinated regions travel continuously, just as in unmyelinated axons. This leads to impairment of the affected motor neuron pathways and is severely debilitating.

Nerve fibres are categorized according to their diameter and action potential conduction speeds. Table 5.1 compares the properties of these fibres. The fastest-conducting fibres (Aα, Aβ, and Aγ) travel to skeletal muscles, where responses are needed very quickly. The slower fibres (Aδ, B, and C) travel from sensory receptors, such as temperature sensing receptors, as the CNS can afford to wait for a short time to receive this information. The diameter of nerve fibres also affects their sensitivity to local anaesthetics, so that the narrowest (and slowest) C fibres are most easily blocked, whereby the thickest (and fastest) A fibres are the last to be blocked. This has useful applications for providing pain relief without blocking muscle power.

> **SELF CHECK 5.4**
>
> Why do the axon diameter and myelination affect the speed of the action potential?

> **SELF CHECK 5.5**
>
> Which nerve fibres conduct the fastest action potentials?

TABLE 5.1 The conduction properties of nerve fibres depend on their diameter and myelination

Fibre type		Myelin	Conduction speed	Diameter	Function
A fibres	Aα	Yes	70–120 m/s	15–20 μm	Primary motor fibres (contraction)
	Aβ	Yes	33–75 m/s	6–12 μm	Motor fibres (muscle stretch detection)
	Aγ	Yes	15–35 m/s	3–6 μm	Secondary motor fibres (**proprioception**)
	Aδ	Moderate	5–30 m/s	1–5 μm	Sensory and **nociceptive** fibres (sharp pain)
B fibres		Moderate	3–15 m/s	1–5 μm	Visceral and autonomic fibres
C fibres		None	<2 m/s	<1.5 μm	Sensory and nociceptive fibres (dull pain)

Adapted from Pocock, G. and Richards, C. *The Human Body*. Oxford University Press, 2009.

5.3 How do nerve cells communicate with each other?

Structure of the peripheral nervous system

Peripheral nerves extend out from the brain and the spinal cord (the CNS) to their target cells throughout the body. Bundles of nerve fibres (known as *nerve trunks*) leave each section of the spinal cord, and extend to different regions. There are two 'roots' for nerves leaving and entering the spinal cord. The nerves that carry signals from the CNS to muscles and glands are known as *efferent motor nerves*, and exit the spinal cord via the *ventral root*. Nerves carrying sensory signals to the CNS are known as *afferent sensory nerves*, and these enter the spinal cord via the *dorsal root*. The efferent nerves that extend to skeletal muscle are known as *somatic nerves* (they are part of the somatic nervous system), while the involuntary nerves, those that supply visceral organs and blood vessels, are **parasympathetic** and **sympathetic** efferent nerves.

 You can learn more about parasympathetic and sympathetic nerves in Section 6.1 'The structure and function of the autonomic nervous system', within Chapter 6 of this book.

The soma of these nerves are bundled together and located according to the direction they carry their signal. The soma of afferent dorsal root nerves (heading toward the CNS) are found in a swelling called the *dorsal root ganglion*, and the cell bodies of efferent ventral root nerves (heading to muscles and glands) lie in the spinal cord itself. In total, 31 pairs of nerve trunks extend from the spinal cord, each carrying many nerve fibres to or from their target cells. As a safeguard, nerve trunks that project from neighbouring parts of the spinal cord usually overlap their targets, so that damage to one nerve trunk does not cause total loss of function in the target organs.

Each peripheral nerve trunk contains the axons of many nerve cells, surrounded by layers of **connec-** tive tissue for protection, and nerves can be up to a metre long. Within this trunk run bundles of axons called *fascicles*, and individual axons may have myelin sheaths. You might like to think of a peripheral nerve as an insulated electrical cable. Inside this insulation are bundles of smaller cables (the fascicles), each containing individual wires (axons), which in turn each carry their own different signal. Take a look at Figure 5.6, which shows a cross section of a peripheral nerve. Efferent motor nerves and afferent sensory nerves usually travel in different nerve trunks, but some nerve trunks carry both motor and sensory fibres, and are called *mixed nerves*. All of the nerves extend from the spinal cord and track the blood vessels closely throughout the body.

Synaptic transmission

Now that we understand how an action potential travels through a nerve fibre, how does it pass from cell to cell? Perhaps surprisingly, nerve terminals do not physically touch their target cells, but the signal is passed by chemical transmission across a narrow gap only 20 nanometres across, called a *synapse*. Take a look at Figure 5.7, which shows the structure of a synapse. The neuron before the synapse is called the

FIGURE 5.6 Diagram showing a cross section through a large nerve trunk. Connective tissue surrounds and protects the fascicles, which contain multiple axon fibres

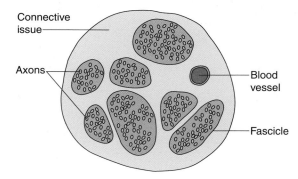

FIGURE 5.7 Diagram of a synapse between a nerve terminal and the target cell. An action potential stimulates release of neurotransmitters from *synaptic vesicles* by exocytosis. The neurotransmitters then diffuse across the synaptic cleft and bind to their target receptors in the postsynaptic cell membrane

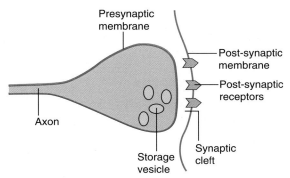

presynaptic cell and the target cell is called the *postsynaptic cell*, which may be another neuron or a muscle or gland cell. At the end of the axon is a small swelling called the nerve terminal, which contains synaptic vesicles (packages) containing molecules called *neurotransmitters*. The postsynaptic membrane on the target cell is richly covered in receptor proteins—the targets for the neurotransmitters. The gap between the two cells of the synapse is called the *synaptic cleft*.

When an action potential reaches the nerve terminal, the presynaptic membrane is depolarized, causing *voltage-gated calcium ion channels* to open. Calcium ions (Ca^{2+}) flow into the presynaptic nerve ending down their electrochemical gradient, and this signals the fusion of synaptic vesicles with the presynaptic membrane. The vesicles then release their neurotransmitters into the synaptic cleft by exocytosis. This entire process is very fast, and happens within 0.5 ms of the action potential arriving. The neurotransmitters diffuse across the synaptic cleft and bind to the target receptors on the postsynaptic membrane of the target cell. Different neurons release different neurotransmitters in response to an action potential at the nerve terminal, but ultimately it is the type of receptors on the postsynaptic cell that determine what happens next in the sequence of events.

 The different types of receptors are covered in Section 4.2 'Basic concepts of pharmacodynamics', within Chapter 4 of this book.

Neurotransmitters involved in synaptic transmission are numerous and diverse, and include the following categories of molecule:

- Esters: *acetylcholine*

- Monoamines: *dopamine, serotonin, noradrenaline*

- Amino acids: *glutamate, γ-aminobutyric acid (GABA), glycine*

- Purines: *adenosine, ATP*

- Peptides: *encephalins, substance P, vasoactive intestinal polypeptide (VIP)*

- Inorganic gases: *nitric oxide.*

KEY POINT

Impulses travel from cell to cell by synaptic transmission, involving chemical messengers called neurotransmitters.

Excitatory and inhibitory synapses

Nerve signals do not all elicit increased activity in a target cell—in fact, some very important signals block or decrease cell activity. The type of receptor on the postsynaptic membrane determines the outcome. If receptor activation on the postsynaptic membrane *increases* activity of the target cell, the synapse is known as an *excitatory synapse*. But if the activation of receptors *decreases* the activity of the target cell, it is an *inhibitory synapse*.

The action on the postsynaptic membrane can also be fast and short, or slow and long, depending on the type of receptor activated. If a *ligand-gated ion channel* is activated, such as with acetylcholine at the neuromuscular junction, synaptic transmission is fast and short, and is called *fast synaptic transmission*. If a **G-protein-coupled receptor** is activated, such as with noradrenaline on α_1-receptors in blood vessels, then transmission is slower and lasts longer, and is called *slow synaptic transmission*.

 You can learn more about G-protein-coupled receptors in Section 4.2 'Basic concepts of pharmacodynamics', within Chapter 4 of this book.

95

Fast synaptic transmission

In *fast excitatory synaptic transmission*, the neurotransmitter released from the presynaptic membrane binds to and opens a *ligand-gated cation channel* on the postsynaptic membrane. Positively charged ions (cations), such as Na^+, enter the cell causing the membrane to depolarize slightly (become more positive), raising the membrane potential from −70 mV closer to the action potential threshold. In neurons, this depolarization is called the *excitatory postsynaptic potential (EPSP)*. EPSPs are usually small (a few mV), peak in less than 5 ms, and fade away within 50 ms. But a single EPSP is too small to raise the membrane potential to the threshold for an action potential. Multiple EPSPs must occur together to sufficiently depolarize the membrane to threshold. This is known as *summation of potentials*—the depolarizing potentials are added together. If the target cell is not a neuron but a skeletal muscle cell then the depolarizing potential is much larger (about 40 mV) and is known as an *endplate potential*. You can read more about endplate potentials in Section 5.4 'Signal transduction in skeletal muscles' within this chapter.

Fast inhibitory synaptic transmission occurs when GABA or glycine are the neurotransmitters in a synapse. These neurotransmitters activate postsynaptic *ligand-gated chloride ion channels*, allowing negatively charged chloride ions (Cl^-) to flow into the postsynaptic cell. This makes the membrane potential slightly *more negative* than −70 mV, known as *hyperpolarization*. This membrane potential change is called an *inhibitory postsynaptic potential (IPSP)*. The membrane potential is now even further away from the threshold potential, making it harder for an action potential to be initiated. Just like the EPSPs, in a fast synapse the IPSP reaches a peak within a few ms and fades away in less than 50 ms.

> **KEY POINT**
>
> Fast *excitatory* synaptic transmission makes action potential generation in the postsynaptic cell *more likely*, whereas fast *inhibitory* synaptic transmission decreases the likelihood of an action potential being initiated.

Neurotransmitter removal

After neurotransmitters have been released during synaptic transmission, there must be a fast mechanism in place to remove the neurotransmitters ready for the next signal. The mechanism depends on the neurotransmitter, but there are three principal ways of halting neurotransmitter action:

- Fast enzymic degradation.
- Uptake into presynaptic cells or neighbouring cells.
- Diffusion out of the synapse before further uptake or enzymatic degradation.

Acetylcholine is the only neurotransmitter to be enzymically degraded by *acetylcholinesterase*, into acetate and choline. These products may be uptaken into the presynaptic nerve terminal and resynthesized into new acetylcholine for reuse. The mono-amine neurotransmitters (such as noradrenaline) are reuptaken from the synaptic cleft into the nerve terminal, where they are stored in vesicles for reuse or destroyed by the enzymes monoamine oxidase or catechol-*O*-methyltransferase. Peptide neurotransmitters diffuse out of synapses and are later destroyed by enzymes, whereas amino acid neurotransmitters are uptaken by neighbouring cells and enter normal metabolism.

 Section 3.3 'Metabolic pathways and abnormal metabolism', within Chapter 3 of this book, looks at the metabolism of amino acids.

Calcium as a second messenger

Calcium is of great importance as an intracellular signal, not only in neurons and muscle but all other cells. The activation of different calcium ion channels by various stimuli allows calcium ions to enter the cell cytoplasm. This rise in intracellular calcium ions is brief, as it is quickly removed back out of the cell or into stores, but this increase in calcium ion concentration acts as a secondary signal in a chain of events that leads to a cellular response, such as contraction of a muscle fibre. In fact, calcium is so important in signalling events that it is often called the *universal* second messenger.

See more about second messengers in Section 4.2 'Basic concepts of pharmacodynamics', within Chapter 4. Integration Box 6.1 within Chapter 6 of this book looks at the role of Ca2+ in the heart.

5.4 Signal transduction in skeletal muscles

If the postsynaptic target cell is a skeletal myocyte (muscle fibre) then the synapse between the end of the signalling neuron and the myocyte is called the *neuromuscular junction*. Here, the neurotransmitter is acetylcholine and the postsynaptic muscle fibre membrane is rich in nicotinic acetylcholine receptors (nAChR). The mechanism of transmission at the neuromuscular junction is similar to that at other synapses. At the postsynaptic cell membrane acetylcholine binds to nAChR, which open and allow sodium ions to enter, depolarizing the myocyte membrane. This depolarization, which is similar to but larger than the EPSP in neurons, is called an *endplate potential (EPP)*. If the EPP is large enough to pass the threshold potential, an action potential is generated in the muscle fibre, which travels along the muscle causing it to contract.

Excitation–contraction coupling

Just like neurons, skeletal muscle cells are excitable and an action potential will quickly travel through the muscle fibre. As an axon approaches a muscle, it branches and each branch connects with a single muscle fibre. The action potential (excitation) spreads along the muscle fibre, causing it to contract. This is known as *excitation–contraction coupling*.

Within myocytes, the sarcoplasmic reticulum contains large calcium ion stores. When the membrane of the myocyte is depolarized, this wave of depolarization spreads along specialized internal extensions of the cell membrane, called T-tubules, to the sarcoplasmic reticulum, causing voltage-gated calcium ion channels to open and calcium ions to enter the sarcoplasm. The rise in sarcoplasmic calcium initiates contraction of the muscle fibre by allowing actin and myosin to interact within the myocyte. Relaxation of the myocyte takes place as the calcium ions are pumped back into the sarcoplasmic reticulum by a calcium ion pump. This demonstrates one very important function of calcium as a second messenger.

5.5 Pain and sensory neurotransmission

Sensory nerves

Receptors in our sense organs (skin, eyes, ears, tongue, and nose) can detect numerous stimuli such as temperature, light, touch, sound, taste, and smell, and the receptors that detect these sensations signal the CNS of these stimuli. These sensory receptors may be part of the afferent neuron or may be separate receptor cells close to the neuron. Receptor activation, by whatever stimulus, results in a change in membrane potential called a *receptor potential*, and the process is known as *sensory transduction*. Receptor activation most commonly causes cation channels to open, depolarizing the nerve membrane slightly (the receptor potential). If the threshold is reached, an action potential is generated at the start of the axon, which then travels along the axon towards the CNS. The intensity of the stimulus depends on the number of receptors activated and the number of action potentials generated. When receptors are first activated after a period of rest, a high-frequency burst of action potentials is generated, but over time this frequency decreases, even if the same stimulus continues. This is known as *sensory adaptation* and explains why we soon accustom to the sensation in our skin of sitting or getting dressed—we soon adapt to the sensation.

Pain

As well as providing the CNS with information about our surroundings, the **somatosensory system** informs the CNS when environmental stimuli become potentially damaging or intolerable, which usually leads to the CNS signalling the body to do something about it, such as move away from the source of this stimulus. The sensations we feel during injury we know as pain—an unpleasant sensation warning us that some-

thing is endangering us. There are three different types of pain sensation we may experience:

- *Sharp pain* is fast and transmitted to the CNS through Aδ fibres.
- *Deep pain* from internal organs or muscles is often difficult to localize.
- *Burning pain* is slow and intense and travels to the CNS via unmyelinated C fibres.

Sensory signals travel more slowly to the CNS than motor signals travel to the skeletal muscles because they use narrower nerve fibres. The slower, narrower fibres are more sensitive to some of the drugs we can use to prevent painful sensations without inhibiting muscle power, such as local anaesthetics.

> **KEY POINT**
>
> Sensory impulses travel to the CNS along slower, thinner nerve fibres than those in motor nerves, and provide information about our external environment.

> **SELF CHECK 5.8**
>
> How does sensory receptor activation lead to a nerve impulse?

5.6 Drugs that affect neurotransmission

We have learned about the numerous important functions that the peripheral nervous system has, and how nerves work and communicate with each other and with other cells. But sometimes we may wish to modify these functions in order to perform medical procedures, or to reduce ongoing pain or alter muscle function. We have a number of drugs available that block or enhance neurotransmission in order to treat various motor disorders, reduce pain, or allow surgery on the body's tissues by preventing muscular contraction and blocking painful sensations. These drugs may

exert their actions at two principal sites: on the neuronal axon itself or at the neuromuscular junction.

Drugs acting on the neuronal axon

Local anaesthetics (*LAs*) are drugs used to numb a localized region of tissue so that pain and sensation in this region are greatly reduced or blocked. LAs do this by acting on the neuronal axon to block the voltage-gated sodium ion channels that are responsible for propagating action potentials. When the sodium ion channels are blocked, action potentials cannot be propagated. Drugs that do this are sometimes called *membrane stabilizing drugs*, as they make depolarizing the cell membrane much more difficult. LAs block the voltage-gated sodium ion channels by crossing the cell membranes then physically plugging the channel from the inside. LAs block more readily when the channels are in the active state, so rapidly firing neurons are more easily blocked. This is called *state-dependent blockade*.

The diameter of neuronal axons affects their sensitivity to LAs. Although all neuronal fibres may be sensitive to LAs, smaller diameter fibres are more sensitive than larger fibres. The smaller diameter unmyelinated C fibres that carry nociceptive signals (pain) are blocked first then, with increasing doses of the LA, thinly myelinated B fibres are blocked and, finally, large diameter myelinated A fibre motor neurons. This size-dependent differential block can be very useful, and with careful dosing allows pain to be blocked without affecting muscle movement. See Table 5.1 for details of the diameter and myelination of nerve fibres.

> **KEY POINT**
>
> Smaller diameter nerve fibres are blocked more readily than larger diameter fibres.

Unlike general anaesthetics, which induce *anaesthesia* (a sleep-like state) through the whole body, local anaesthetics are so named because of their ability to work in localized areas when given in an appropriate dosage form. An ointment may contain an LA, an injection may be given for a dental procedure, or LAs may even be injected for spinal or epidural anaesthesia to block entire nerve tracts. Case Study 5.1 shows some of the different ways LAs can be beneficial and Box 5.4 introduces some effects of neurotoxins.

> **KEY POINT**
>
> Local anaesthetics block voltage-gated sodium ion channels in neuronal axon membranes, preventing the propagation of action potentials.

CASE STUDY 5.1

Rachel went to her dentist to have a filling and was given a local anaesthetic which caused the left side of her face to go completely numb and she couldn't smile or feel her cheek for over an hour afterwards. Rachel had also recently received a local anaesthetic as an epidural while giving birth to her baby, which numbed the pain of childbirth but she could still feel the sensation of birth and move her legs.

REFLECTION QUESTIONS

1. Why didn't the dentist give Rachel painkillers instead of the local anaesthetic?

2. Why did Rachel have different sensory effects from the two local anaesthetics she received at the dentist and during childbirth?

3. Careful dosing with local anaesthetic drugs can be used to bring about analgesia, anaesthesia, and temporary muscle paralysis. But what is the difference between these three states?

Answers

1 Painkillers, or analgesics as they are properly called, will help to reduce pain but are unlikely to completely prevent the pain that would be felt during a surgical operation, like having a tooth drilled. The dentist injected a local anaesthetic drug to numb the region around the tooth completely, so no pain or sensation at all would be felt during the procedure.

2 Local anaesthetic drugs can have different effects when dosed very carefully in different parts of the body. Lower doses of anaesthetics injected into the epidural space around the spine will block smaller diameter nerves carrying painful signals without completely blocking the larger diameter nerves that carry sensations and muscle signals. Doses injected around the inferior alveolar nerve in the lower jaw block the sensation in the nerve and result in the loss of feeling in and around the gums, lip, and chin.

3 Analgesia is relief from pain, anaesthesia is loss of sensations such as touch and temperature, including pain, and paralysis is loss of muscle power. Whilst muscle paralysis requires a relatively high dose of local anaesthetic drug to block the larger diameter neurons, analgesia and anaesthesia can be achieved with lower doses.

Sodium ion channel blockers: toxic fine dining

Nature provides a number of compounds that block the voltage-gated sodium ion channels of nerve axons. The pufferfish contains deadly tetrodotoxin, which it accumulates from symbiotic bacteria living in its gut. Tetrodotoxin is a potent blocker of voltage-gated sodium ion channels and can cause fatal blocking of nerve and muscle (including heart) action potentials in humans. Pufferfish are considered the second most poisonous vertebrate in the world, but highly trained Japanese chefs serve some less toxic parts of the fish to diners who consider it a delicacy. Sensations from eating the fish include tingling and numbness of the tongue and lips, but the effect is only temporary. Tetrodotoxin has also been used medicinally in some countries for its analgesic effect, and may be useful in treating some types of cancer pain. In fact, many other neurotoxins from naturally occurring venoms and plant alkaloids offer us some potentially very effective and potent treatments when doses are carefully controlled.

How do local anaesthetics work?

Drugs acting at the neuromuscular junction

Another important site of action for drugs to alter neurotransmission is at the neuromuscular junction. Here, drugs are used not to control pain but to affect the transmission of signals to muscle fibres. They can be categorized as:

- Drugs that *block* neurotransmission by directly targeting the postsynaptic receptors at the junction.

- Drugs that *enhance* neurotransmission by altering levels of neurotransmitters at the junction.

Neurotransmission blocking drugs

During complex surgical procedures, delicate tasks require the patient to be very still. As well as receiving an anaesthetic drug to prevent pain during the operation, a patient will often also be given a *neuromuscular blocking drug* to induce temporary muscle paralysis. Neuromuscular blocking drugs can be described according to their mode of action, as *non-depolarizing blockers* or *depolarizing blockers*.

Non-depolarizing blockers: One of the earliest identified compounds that caused muscle paralysis was *curare*—a poisonous plant alkaloid. This drug has been used for centuries by South American tribesmen to coat their arrows, so that during hunting they can dart animals and cause respiratory muscle paralysis, making prey easier to catch. In medicine we use neuromuscular blocking drugs such as tubocurarine, a synthetic curare derivative, in surgical procedures for the same effect—relaxation of muscle. Drugs such as curare and tubocurarine work by binding to and blocking the postsynaptic nAChR at the neuromuscular junction, so that the postsynaptic cell cannot be depolarized and the action potential signal cannot be transmitted to the muscle. For this reason they are called *non-depolarizing blockers*. Other examples of these drugs in clinical use include atracurium, pancuronium, and vecuronium. These drugs have an effect on all voluntary skeletal muscles, including the diaphragm, so muscles become paralysed and breathing stops. Although useful for preventing unwanted muscle movements, patients must be artificially ventilated so they can continue to breathe until the drug effects wear off.

Depolarizing blockers: This group of drugs also causes neuromuscular block, but by a different mechanism. Rather than simply blocking nAChR, depolarizing blockers mimic the action of acetylcholine on the nAChR of the neuromuscular junction, causing depolarization of the postsynaptic cell. But, in contrast to acetylcholine, depolarizing blockers occupy the receptor for much longer and continuously trigger depolarization, maintaining the membrane potential above threshold. Therefore acetylcholine itself cannot act to depolarize the muscle cell. When a depolarizing drug, such as suxamethonium, is applied, two phases of block are observed. During phase 1 (depolarization), initial twitching contractions of the muscle occur, called *fasciculations*, as the drug depolarizes the muscle causing contractions. But after a few seconds phase 2 occurs (desensitization), in which there is a loss of electrical excitation as voltage-gated sodium ion channels remain in their inactive state and are therefore unable to reopen with continued stimulation. The muscle is now paralysed in a relaxed state, blocked from further depolarization.

Suxamethonium is the only depolarizing blocker used clinically. It acts quickly and is broken down at the neuromuscular junction by a cholinesterase enzyme (butyrylcholinesterase) that takes several minutes to act. This makes it a very useful drug for clinical use to relax tracheal muscle when patients are being intubated to aid breathing.

> ### KEY POINT
>
> Non-depolarizing blockers are **antagonists** of nicotinic acetylcholine receptors at the neuromuscular junction, causing neuromuscular block. Depolarizing blockers *activate and then desensitize* nicotinic acetylcholine receptors to cause neuromuscular block.

Neurotransmission-enhancing drugs

Another way to manipulate neurotransmission at the neuromuscular junction is by increasing the availability of neurotransmitters, thereby affecting their action on the postsynaptic receptors of the muscle fibre. These drugs act to keep neurotransmitters in the synaptic cleft for longer and may therefore enhance neurotransmission. *Acetylcholinesterase inhibitors*, such as neostigmine and physostigmine, act by inhibiting acetylcholinesterase, which naturally degrades acetylcholine in the synapse and is important for allowing discrete signalling by neurons, thereby allowing muscle to relax. By blocking this enzyme, acetylcholine remains in the synaptic cleft for longer, and has a prolonged action on the postsynaptic receptors.

One disease in which acetylcholinesterase inhibitors are useful is myasthenia gravis. People with this condition have **antibodies** which target and destroy the body's own nAChR at the neuromuscular junction. This leads to impaired neuromuscular transmission and muscular disability. By allowing acetylcholine to act for longer on the remaining nAChR, neuromuscular transmission can be improved, helping to reduce the disabling effects of the disease.

However, *high* doses of acetylcholinesterase inhibitors can cause muscle paralysis owing to the resultant prolonged levels of acetylcholine at the neuromuscular junction. This is depolarizing blockade, with a mechanism similar to that of drugs like suxamethonium. Many other compounds exist which interfere with neuromuscular transmission, and Box 5.5 considers one of them.

> ### SELF CHECK 5.10
>
> Describe the two possible mechanisms for neuromuscular block.

> ### BOX 5.5
>
> ## Neurotoxins in medicine
>
> Neurotoxins may occur naturally in venoms or as plant alkaloids, and have been used as chemical weapons in warfare. Botulinum toxin, produced by the bacterium *Clostridium botulinum*, is one of the most potent neurotoxins known. It can be ingested in contaminated foods, causing botulism, which results in severe muscle paralysis by preventing acetylcholine being released from nerve terminals, inhibiting neurotransmission. But once again, modern medicine has found a medical use for this toxin in treating spasticity (muscle rigidity) in some motor neuron diseases, and in reducing the appearance of wrinkles by paralysing facial muscles in cosmetic surgery—you may know this drug by one of its brand names: Botox®.

CHAPTER SUMMARY

➤ The peripheral nervous system carries rapid and discrete electrical signals between cells, often over long distances.

➤ The somatic division of the peripheral nervous system carries sensory signals (including pain) to the central nervous system, and motor signals to skeletal muscle.

➤ Neurons are excitable cells that carry electrical signals called action potentials along their axons.

➤ An action potential occurs when a sensory signal causes the membrane potential of a neuron to become more positive (depolarize).

➤ An action potential propagates along an axon owing to the opening of voltage-gated sodium ion channels along the axon, which causes the axon membrane to depolarize.

➤ Neuroglia (such as Schwann cells) support the role of the neurons and form the myelin sheath around many nerve axons.

➤ The speed of action potentials depends on the diameter of the axons, and on myelin coatings: thicker, myelinated axons carry faster action potentials.

➤ Neurotransmission occurs from cell to cell across narrow gaps called synapses, using neurotransmitter molecules to carry the signal. A synapse between a neuron and a muscle fibre is called a neuromuscular junction.

➤ An action potential (excitation) that causes muscle contraction is known as excitation–contraction coupling.

➤ Drugs that act on the somatic part of the peripheral nervous system can be useful to prevent pain or relax muscles during surgery.

➤ Local anaesthetics block voltage-gated sodium ion channels on nerve axons, and smaller diameter axons (carrying pain and sensory signals) are more sensitive than larger diameter motor neurons.

➤ Neuromuscular blocking drugs relax muscles by preventing neurotransmission of signals for contraction, which can be useful during surgery. They work by inhibiting activation of postsynaptic nicotinic acetylcholine receptors at the neuromuscular junction.

➤ Some drugs can act to enhance neurotransmission by elevating or prolonging levels of neurotransmitters. These work by inhibiting the cholinesterase enzyme that breaks down acetylcholine at the neuromuscular junction.

FURTHER READING

Rang, H.P., Dale, M.M., Ritter, J.M., Flower, R.J., and Henderson, G. *Rang and Dale's Pharmacology.* 7th edn. Churchill Livingstone, 2011.

Section 4 of this book, entitled *The nervous system,* presents further information on neurotransmission and neurodegenerative diseases, with more detailed descriptions of the pharmacology of local anaesthetics and neuromuscular blockers.

Pocock, G. and Richards, C. *Human Physiology: The Basis of Medicine.* 3rd edn. Oxford University Press, 2009.

This textbook presents a much more detailed account of the physiology and anatomy of the peripheral nervous system, including somatosensory and motor connections and control.

The autonomic nervous system

EMMA LANE AND WILLIAM FORD

Think about what happens to your body during an emergency—that 'fight or flight' situation: your heart beats faster, your blood is pumping, you might sweat, you feel the adrenaline surge and you are ready for action!

Not only does your body start doing all these things but it stops actions that are not useful in these situations, such as digesting food and producing saliva. These bodily functions, either in moments of activity or rest, are controlled by part of the nervous system known as the autonomic nervous system (ANS). This chapter deals with the basic physiology of the ANS, the functions it provides to the different organs, and the neurochemical transmission that occurs within it, and highlights the uses of pharmacotherapy in its modulation. Although not under voluntary control, the autonomic system does not work independently of the central nervous system. Specific brain regions supervise and influence the functions of the ANS and this flow of information and control will also be described.

Learning objectives

Having read this chapter you are expected to be able to:

➤ Compare neurochemical transmission in the sympathetic and parasympathetic nervous systems.

➤ Appreciate how these two systems complement each other and are influenced by the central nervous system.

➤ Describe the functions of the sympathetic and parasympathetic nervous systems in a variety of organ systems.

6.1 The structure and function of the autonomic nervous system

The nervous system, as described in Chapter 5, is made up of several components, each defined by a specific set of functions and with a distinct anatomy and physiology. As an integrated part of the periph-eral nervous system, the autonomic nervous system (ANS) discretely, and without conscious thought, controls the visceral functions of the body which are fundamental to sustaining life. Figure 6.1 shows the

FIGURE 6.1 The autonomic nervous system is a subsection of the nervous system controlling the involuntary functioning of many internal organs. It can be considered in two divisions, the sympathetic and parasympathetic nervous systems, working alongside the enteric nervous system

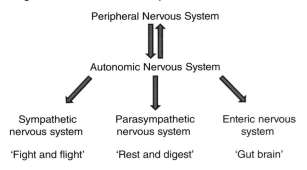

two main divisions of the ANS, the *parasympathetic* and *sympathetic* nervous systems, which can broadly be described as controlling the 'rest and digest' and 'fight or flight' responses, respectively. This is an over-simplification but these descriptors serve a purpose in illustrating the responsibilities of the ANS in maintaining homeostasis within the body and responding to changes in the environment.

The sympathetic nervous system can be seen as readying the body for action, accelerating heart rate and slowing functions that are not necessary in moments of stress or activity. Conversely, the parasympathetic nervous system slows the heart and promotes actions such as digestion, micturition, and bile secretion. These complementary systems are constantly working to fine-tune the control of many bodily functions. Figure 6.1 also identifies the *enteric nervous system*, a small but vital system within the ANS often referred to as the 'gut brain'. It is made up of two types of ganglia that, independently of the rest of the nervous system, control the functioning of the gut.

The anatomy and physiology of the autonomic nervous system

The function of the ANS is to regulate vascular and visceral smooth muscle, exocrine and some endocrine functions, the heart, and energy metabolism (particularly in the liver and skeletal muscle). It provides a basal level of activity in the tissues it innervates, which is referred to as *tone. Sympathetic tone*, for example,

maintains a constant partial contraction of the blood vessels, which means that a certain level of blood pressure is maintained. From this baseline, increasing or decreasing the sympathetic activity will have a direct consequence, raising or lowering blood pressure, respectively. The heart, on the other hand, is predominantly regulated by the vagus nerve of the parasympathetic nervous system, providing *vagal tone*. If the slowing vagus nerve is severed, the heart beats faster.

Generally, the sympathetic and parasympathetic nervous systems have opposing functions but there are exceptions, such as the salivary and lacrimal glands, where both systems support the same function.

The anatomical arrangements of the sympathetic and parasympathetic branches are different, and they differ again from the somatic motor innervations, as shown in Figure 6.2. In the somatic nervous system, spinal motor neurons directly synapse with skeletal muscle using acetylcholine as the neurotransmitter. However, innervation of target organs by the ANS involves a two-neuron pathway via *ganglia*. Ganglia (singular = ganglion) are collections of neuron cell bodies bundled together. Neurons connecting the spinal cord to these ganglia are termed *preganglionic* fibres or neurons whereas neurons that synapse at the ganglia and travel out to the effector target organs are known as *postganglionic* fibres.

 You can learn more about the somatic nervous system in Chapter 5 of this book.

Neurons within the ANS release one of two neurotransmitters: acetylcholine or noradrenaline. *Cholinergic*

FIGURE 6.2 The characteristic neuroanatomy of the somatic, sympathetic, and parasympathetic nervous systems

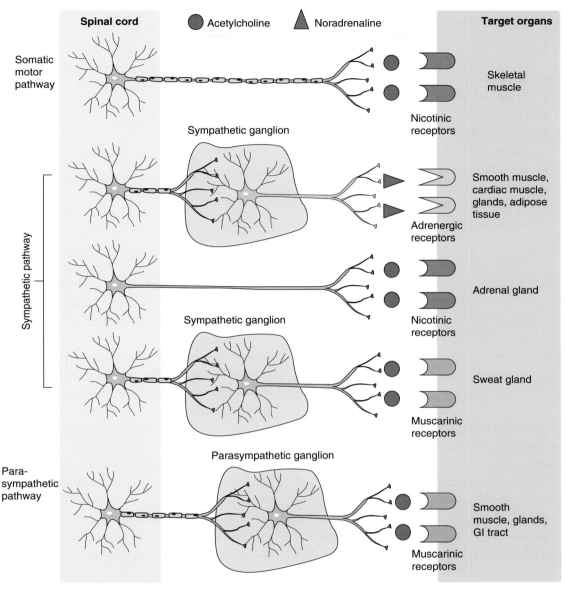

neurons release acetylcholine while those that release noradrenaline are called *noradrenergic neurons*. All **efferent** neurons that emerge directly from the spinal cord are cholinergic and myelinated, whereas postganglionic neurons can be both cholinergic and noradrenergic. In the parasympathetic nervous system both the preganglionic and the postganglionic neurons are cholinergic. In the sympathetic nervous system, the preganglionic neurons are cholinergic but the postganglionic are noradrenergic. There are, however, two exceptions to the neuronal anatomy described for the sympathetic pathway, as shown in Figure 6.2:

• In the **adrenal medulla** cholinergic preganglionic fibres pass to the adrenal glands via the *splanchnic nerves* and synapse directly onto the **chromaffin cells**, which then trigger the release of adrenaline and noradrenaline into the bloodstream. This systemic release of the sympathetic nervous system neurotransmitters adds to the specific neurotransmission at each sympathetic synapse and acts on relevant receptors at other locations.

 Section 4.2 'Basic concepts of pharmacodynamics', within Chapter 4 of this book, looks into receptors and the different types in more detail.

- Sweat glands are also an exception in the sympathetic nervous system, as the postganglionic neurons are cholinergic rather than noradrenergic. They still form part of the sympathetic nervous system, however, because the pre- and postganglionic neurons synapse in the superior **cervical** ganglion at the top of the sympathetic trunk, as explained under 'The sympathetic nervous system' within this section.

> **KEY POINT**
>
> The neurotransmitters released and their associated receptors are involved in distinguishing the responses within the autonomic nervous system from those of the somatic pathway.

The sympathetic nervous system

In the widely innervating sympathetic nervous system, preganglionic neurons originate from the **thoracic** and top **lumbar** regions of the spinal cord (T1 to L2, see Box 6.1 and Figure 6.3). Alongside the somatic motor fibres, the preganglionic fibres pass from the **lateral horn** of the spinal cord via the **ventral spinal root**, travelling a short distance to the sympathetic ganglia. These ganglia are located *paravertebrally* in

> **BOX 6.1**
>
> ## The spinal segments
>
> The spinal cord is protected by the bony vertebral column, which is made up of 31 spinal segments that are subdivided into sections. C1 to C7 are the cervical segments at the top of the spinal column, followed by T1 to T12, the thoracic region, then L1 to L5 the lumbar region and finally the S1 to S5 sacral region. There are also three segments that fuse to form the coccyx. One pair of nerves exits from each spinal segment.

the *sympathetic trunk* found proximal to either side of the spinal cord. The **myelinated** nerve axons of the cholinergic preganglionic fibres lie within the white **rami communicantes** (myelination gives the white appearance). Upon reaching the ganglia, the preganglionic fibres synapse with postganglionic noradrenergic fibres, which then project a long distance to the target organs via the grey rami communicantes (grey due to the lack of myelination).

The parasympathetic nervous system

The parasympathetic nervous system contains much longer cholinergic preganglionic fibres than the sympathetic nervous system and again it originates in the spinal cord (as shown in Figure 6.2), but this time fibres emerge at the level of the brain stem and in **sacral** segments S2 to S4 without passing through rami communicantes. Four **cranial** nerves emerge from the brain stem: III (oculomotor), VII (facial), IX (glossopharyngeal), and X (vagus). These contain many sensory **afferent** neurons as well as somatic and parasympathetic efferent neurons. The ganglia lie close to or within the target organ, such that only short cholinergic postganglionic neurons are necessary.

In its capacity for controlling the 'rest and digest' processes, parasympathetic activation constricts pupils (via the oculomotor nerve); secretes tears (via the facial nerve), saliva (via the glossopharyngeal nerve), and bile; initiates micturition; increases gut secretion and motility; decreases heart rate; constricts bronchi; and dilates a select number of blood vessels (all via the vagus nerve). As you can see from this list, and Figure 6.3, the vagus nerve is a major regulator of the parasympathetic nervous system. It performs these functions by direct innervation of the heart, as well as the trachea and lungs, liver, stomach, small and large intestines, and the kidney. The lower sacral contribution to the parasympathetic nervous system innervates the large intestine, bladder, and genitalia.

> **KEY POINT**
>
> There are anatomical differences between the sympathetic and parasympathetic nervous systems, which are summarized in Table 6.1.

The item below is now available for pickup at designated location.

Grace A Dow Mem Library
zv225 Grace A Dow Mem Library

Baker - Allen Park
CALL NO: RM300 .T44 2013
AUTHOR:
Therapeutics and human physiology :
BARCODE: 33504005823721
REC NO: i403316972
PICKUP AT: Grace A Dow Mem Lib

MARIA C CASEY
Grace A Dow Mem Library

989-859-7390

FIGURE 6.3 The innervation of different visceral organs by the two parts of the autonomic nervous system. Note the location of the sympathetic trunk next to the spinal cord. Only a few postganglionic fibres are shown for the parasympathetic nervous system as the majority of ganglia are immediately next to or within the target organ

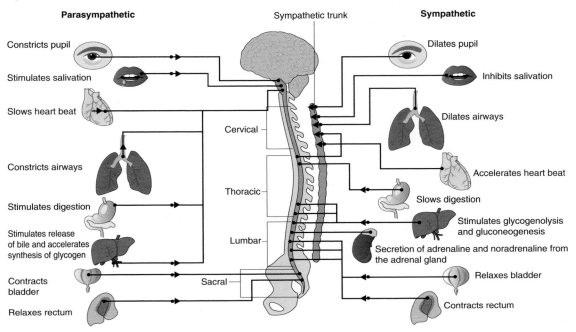

The enteric nervous system

The enteric nervous system is a small and distinct part of the ANS that controls gut function. Its direct connections with the central nervous system (CNS), its complexity, high number of neurons, and autonomy of function have earned it the name of 'second brain' or 'gut brain'. The enteric nervous system is responsible for the control of the complex propulsive gut peristaltic motions and intestinal blood supply, along with secretions.

The enteric nervous system has its own neuronal anatomy: two main ganglionic elements act to control the smooth muscle wall of the gut and the secretory epithelium between the layers of muscle and the lumen of the gut. The *myenteric plexus* (also known as *Auerbach's plexus*) largely controls motor function

TABLE 6.1 Differences between the anatomy of the sympathetic and parasympathetic nervous systems

	Sympathetic	Parasympathetic
Ganglia	Proximal to spinal cord, distal to effector	Distal to spinal cord, proximal to effector
Emerging at spinal cord levels	Thoracic and lumbar (T1 to L2)	Cranial and sacral (S2 to S4)
Preganglionic fibres	Short Myelinated, cholinergic	Long Myelinated, cholinergic
Postganglionic fibres	Long Unmyelinated, noradrenergic	Short Unmyelinated, cholinergic
Rami communicantes	Grey and white	No rami

and lies between the layers of smooth and circular muscle of the gut wall, while the *submucosal plexus* (also known as *Meissner's plexus*) controls gut secretion and blood flow to the gut, and is located within the submucosal layer (shown in Figure 6.4). Within each plexus the cell bodies of the neurons are organized into ganglia, which innervate each other as well as the blood vessels, smooth muscle and glands of the gut.

Autonomic reflexes and central control of the ANS

The independence of the sympathetic and parasympathetic branches of the ANS does not mean that they are unregulated. The organs controlled by the ANS are innervated by afferent fibres, which respond to mechanical and chemical changes through baroreceptors, mechanoreceptors, and chemoreceptors. The ANS functions in part through reflexes mediated at the level of the spinal cord or brain stem. Examples of these reflexes in action are lung inflation, covered in Section 6.5, and micturition, covered in Section 6.9.

The maintenance of homeostasis—a constant environment optimal for the body to function—is primarily dealt with by the *hypothalamus*, a small structure located at the base of the brain. The hypothalamus can be subdivided into different nuclei

and the most critical for autonomic control is the *paraventricular nucleus* (PVN). Inputs to the PVN are received from somatic sensory and visceral afferent nerves, allowing immediate responses to changes in the internal and external environments. Direct connections are formed between the PVN and the spinal cord, brain stem, and vagus nerve. Direct neuronal communication, as well as the release of regulatory hormones, gives this brain region overarching control of both the sympathetic and parasympathetic nervous systems.

The brain stem, the anatomical connection between the spinal cord and the rest of the brain, contains many sites which regulate the sympathetic outflow. There are also extra-hypothalamic brain regions that influence parasympathetic outflow. The parts of the brain that deal with pain and emotion, the cerebrum and limbic system respectively, also communicate with the hypothalamus, explaining how emotional responses can trigger physiological effects such as altered heart rate and blood pressure.

> **KEY POINT**
>
> The spinal cord and autonomic ganglia are the major sites for autonomic control but there is also significant communication and influence from higher centres of the CNS, for example the hypothalamus and cerebrum.

FIGURE 6.4 A transverse section of the gut showing the location of the myenteric plexus and submucosal plexus in different layers of the gut wall
Source: Reproduced with permission, from Tiffany A. Heanue & Vassilis Pachnis (2007) *Enteric nervous system development and Hirschsprung's disease: advances in genetic and stem cell studies.* Nature Reviews Neuroscience **8**, 466–479

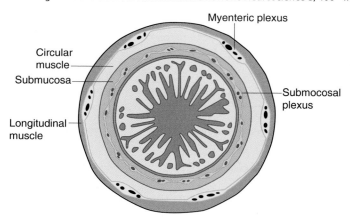

6.2 Neurotransmission within the ANS

As illustrated in Table 6.2, many organs receive innervation from both the sympathetic and parasympathetic nervous systems, often working in opposition. In some organs the same neurotransmitter has opposing effects owing to the presence of different receptor subtypes for that neurotransmitter. An example of this is the effects caused by activation of the α- and β-adrenoceptors in the vascular smooth muscle described in Section 6.4 'Effect of the ANS on the cardiovascular system'.

There are two predominant neurotransmitters of the ANS, acetylcholine and noradrenaline, each of which stimulates distinct receptors and the action of which is often in opposition. We will consider each of these in turn.

TABLE 6.2 The action of the sympathetic and parasympathetic nervous systems and the postsynaptic receptor(s) that mediate their function

Body system	Organ/muscle	Sympathetic		Parasympathetic	
		Receptor	Action	Receptor	Action
Cardiovascular	Heart	β_1 (and β_2)	Increased rate and force of contraction	M_2	Decreased heart rate
	Coronary blood vessels	α_1	Vasoconstriction	M_3	Vasodilatation
Respiratory	Smooth muscles of bronchioles	β_1	Dilates bronchioles	M_3	Constricts bronchioles
	Blood vessels	α_1	Vasoconstriction	M_3	Vasodilatation
Oculomotor	Pupil dilator muscle	α_1	Contraction (dilation)		
	Iris sphincter muscle			M_3	Contraction (constriction)
	Ciliary muscle	β_2	Relaxation	M_3	Contraction
Hepatic	Liver	α_1/β_2	Promotes glycogenolysis	M_3	Glucose homeostasis
	Gall bladder	α_2	Inhibits bile release	M_3	Stimulates bile release
Digestive	Digestive tract	α_1/β_2	Decreased digestion	M_1/M_3	Increased digestion
Renal and urinary	Kidney	α_1	Reduced urine output		
	Detrusor muscle	β_2	Bladder relaxation	M_3	Bladder contraction
	Internal sphincter	α_1	Urethral contraction	M_3	Urethral relaxation
Endocrine	Pancreas	α_2	Glucagon release	M_3	Insulin release
	Adrenal medulla	N	Noradrenaline and adrenaline release		
Reproductive	Uterus	α_1	Contraction (pregnant)		
		β_2	Relaxation (not pregnant)		
	Genitalia male	α_1	Ejaculation	M_3	Erection
	Genitalia female	α_1/α_2	Vaginal contraction and mucus secretion	M_3	Erection
Integumentary	Sweat glands	M_4	Increased sweating		
	Arrector pili muscles	α_1	Contraction ('goose bumps')		

α, α-adrenoceptor; β, β-adrenoceptor; N, nicotinic receptor; M, muscarinic receptor.

Acetylcholine and cholinergic receptors

Acetylcholine is released by the preganglionic nerves in the sympathetic nervous system and throughout the parasympathetic nervous system. Acetylcholine is formed in the presynaptic terminal by the enzymatic action of *choline O-acetyltransferase*, which removes an acetyl group from acetyl coenzyme A and links it to choline, an essential dietary nutrient released by the mitochondria. The fully formed acetylcholine is then packaged into vesicles ready for release by the postsynaptic membrane when an **action potential** is received. *Acetylcholinesterases* are enzymes located in the synaptic cleft that rapidly break down the released acetylcholine back into its constituent parts. As the acetyl group diffuses away, choline is actively taken back up into the cell by a transporter on the presynaptic membrane and recycled, as shown in Figure 6.5.

 The formation of action potentials is described in Section 5.2 'An exciting journey: how are signals transmitted?', within Chapter 5 of this book.

The receptors responsive to acetylcholine are the cholinergic receptors, of which there are two types: *nicotinic* and *muscarinic*. Nicotinic receptors are **ligand-gated ion channels** located at the cholinergic neuromuscular junction of the somatic motor system and at the autonomic ganglia. However, the target organs of the parasympathetic nervous system express muscarinic receptors that are a type of **G-protein-coupled receptor**. The consequences of the neurotransmitter binding to the G-protein-coupled receptor on the cellular processes can be stimulatory or inhibitory, depending on the type of G-protein the receptor associates with.

 The different types of receptor are covered in more detail in Section 4.2 'Basic concepts of pharmacodynamics', within Chapter 4 of this book.

There are five distinct subtypes of muscarinic receptor:

- M_1 receptors are generally found on neurons in the CNS but are also in peripheral ganglia. They control gastric secretions and are associated with an excitatory G-protein (G_q).

- M_2 receptors are located in the heart and are inhibitory (G_i).

FIGURE 6.5 Acetylcholine synthesis and release from cholinergic synapses, found at the ends of motor neurons and at autonomic ganglia

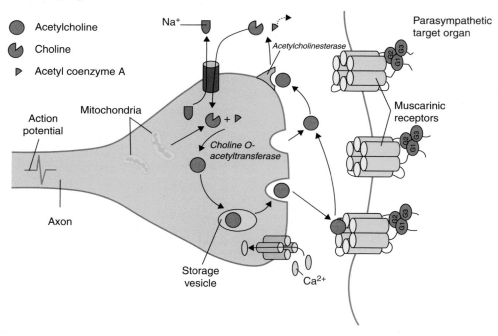

- *M₃ receptors* are the predominant cholinergic receptor subtype at the autonomic target organs, mainly secretory glands and smooth muscle, and are linked to an excitatory G-protein (G_q).

- *M₄ and M₅ receptors* are more restricted to the CNS and are less well characterized, M_4 being inhibitory (G_i) and M_5 excitatory (G_q). They are also located on salivary glands and the parasympathetic innervation of the eye.

Noradrenaline and adrenergic receptors

Noradrenaline (also referred to as norepinephrine) is released by postganglionic neurons in the sympathetic nervous system. Noradrenaline is formed in the presynaptic terminal from tyrosine, a non-essential amino acid obtained from many foods or synthesized from phenylalanine. Once taken up into the cell it is converted, in a rate-limiting step, by the enzyme *tyrosine hydroxylase* into L-dihydroxyphenylalanine (L-DOPA) before the *amino acid decarboxylase* enzyme converts it into dopamine. Dopamine is then transported into the vesicles, where *dopamine β-hydroxylase* completes the transition to noradrenaline. This process is shown in Figure 6.6. On release of noradrenaline into the synaptic cleft, specific reuptake transport proteins in the pre- and postsynaptic cell membranes transport the noradrenaline back into the pre- or postsynaptic cell. The noradrenaline is then either repackaged back into vesicles for release or broken down by enzymes, *catechol O-methyltransferase* and *monoamine oxidase*.

Adrenaline (also referred to as epinephrine) is chemically related to noradrenaline and they both belong to the catecholamine class of chemicals. However, instead of being released by nerves and acting as a neurotransmitter, adrenaline is released by the adrenal medulla into the blood. It acts as a *hormone,* causing effects at a distance from where it is released, but it stimulates the same receptors within the sympathetic nervous system as noradrenaline so it can have the same effects.

The adrenergic receptors, often referred to as *adrenoceptors,* fall into two categories, α and β, with subtypes α_1, α_2, β_1, β_2, and β_3. All of these receptors are G-protein-coupled receptors and are found

FIGURE 6.6 Synthesis and release of noradrenaline from noradrenergic synapses found selectively at target organs of the sympathetic nervous system

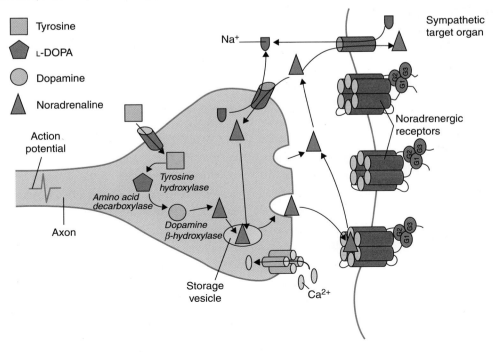

throughout the CNS as well as the autonomic system (see Table 6.2):

- α_1-adrenoceptors are found in much of the smooth muscle innervated by the noradrenergic sympathetic system. Released adrenaline also activates these receptors, which are linked to excitatory G-proteins (G_q).

- α_2-adrenoceptors are inhibitory but their stimulation also causes contraction of the blood vessels that express them. These receptors are also located presynaptically and act as negative feedback regulators of neuronal firing, inhibiting noradrenaline release from the terminals (G_i).

- β_1-adrenoceptors are found in the heart, where they have an excitatory effect (G_s), although in the gut their activation leads to relaxation (also through G_s because of differences in the way cardiac and smooth muscle contract).

- β_2-adrenoceptors are expressed in smooth muscle and some blood vessels, leading to vasodilatation when activated, and in bronchial smooth muscle, where activation results in bronchodilation (G_s).

- β_3-adrenoceptors are present in fatty adipose tissue and some smooth muscle (intestine, gall bladder, and urinary bladder) (G_s). β_3-adrenoreceptor function is unclear but has been proposed to mediate smooth muscle relaxation in the tissues in which it is present.

6.3 Drug action in the ANS

Now that we understand the basics of neurotransmitter release and the receptors within the ANS they act on, we can consider how therapeutic molecules may interfere with synaptic transmission.

Direct-acting therapeutic molecules

The response caused by the activation of the postsynaptic receptor may be enhanced or prevented by the administration of an agonist or an antagonist, respectively. Drugs acting as agonists will stimulate the receptors and cause the subsequent physiological response. *Parasympathomimetics* mimic the effects of acetylcholine at the parasympathetic innervation. *Sympathomimetics*, however, replicate the noradrenergic activity of the postganglionic sympathetic nervous system. Pilocarpine is an example of a parasympathomimetic used in the treatment of glaucoma, whereas the sympathomimetic salbutamol is used in the management of asthma.

Antagonists will block access of the relevant endogenous neurotransmitter to the receptor. This prevents the respective physiological change from occurring. Direct-acting *parasympatholytics* block or reduce the action of acetylcholine. On the other hand, direct-acting *sympatholytics* compete with noradrenaline and adrenaline at the adrenergic receptors. Ipratropium is a parasympatholytic that is used as a bronchodilator in the management of asthma and chronic obstructive pulmonary disorder. Prazosin is an example of a sympatholytic used for hypertension. See Table 6.3 for other clinically relevant examples of direct-acting agonists and antagonists of the ANS.

Indirect-acting therapeutic molecules

Prolonging the availability of the neurotransmitter in the synaptic cleft is another way of achieving a higher synaptic concentration of the relevant endogenous molecule. This can be achieved in a number of different ways and is dependent on the neurotransmitter. In the case of noradrenaline, reuptake of the transmitter presynaptically or postsynaptically can be blocked with reuptake inhibitors, whereas the same effect is produced in cholinergic synapses if the metabolizing enzyme, acetylcholinesterase, is blocked by acetylcholinesterase inhibitors. Methylphenidate is an example of an indirect sympathomimetic, inhibiting noradrenaline

TABLE 6.3 Clinically relevant categories of autonomic drugs, their uses, and associated side effects

Drug type	Action	Examples	Uses	Side effects
Parasympathomimetic agents	Muscarinic agonist	Pilocarpine	Little medical use, glaucoma, post-operative urinary retention	Bronchospasm, asthma, nausea, vomiting, cramps, diarrhoea
	Acetylcholinesterase inhibitor	Captopril Enalapril	Laxative, hypertension, heart disease and heart failure, diabetic nephropathy	Excessive secretions, GI cramps, diarrhoea, bradycardia
Parasympatholytic agents	Muscarinic antagonist	Ipratropium Tiotropium Hyoscine	Asthma[†], bronchitis, COPD, menstrual or GI cramps, motion sickness, excessive secretions	Dry mouth/skin, blurred vision, tachycardia, flushed face, increased body temperature, significant CNS effects
Sympathomimetic agents	Non-selective adrenergic agonist	Adrenaline Ephedrine	Shock, cardiac stimulation Nasal decongestion	Tachyarrythmias, palpitations, hypertension, CNS stimulation
	α_2-adrenoceptor agonist	Clonidine	Asthma, cardiac stimulation, hypotension, menopausal flushing	
	β_2-adrenoceptor agonist	Salbutamol (β_2, short-acting) Salmeterol (β_2, long-acting)	Hypotension, asthma, bronchospasm, cardiac shock, postpartum bleeds, mydriasis, premature labour	
Sympatholytic agents	Non-selective adrenoceptor antagonist	Carvedilol Labetalol	Hypertension, heart failure	Low blood pressure, tachycardia, nasal congestion
	α_1-adrenoceptor antagonist	Prazosin Doxazosin Tamsulosin Phenoxybenzamine	Hypertension Benign prostatic hyperplasia[‡] Phaeochromocytoma[§]	Orthostatic hypotension, nasal congestion, urinary urgency
	β-adrenoceptor antagonist	Propranolol Atenolol Timolol	Hypertension, cardiac arrhythmias, angina, myocardial infarction, glaucoma	Bradycardia, bronchoconstriction, hypotension, cardiac depression, impotence

Some of these drugs also have CNS actions and may be used differently for CNS conditions, but the side effects will be the same. This list is not comprehensive and the drugs named are only exemplars.

CNS, central nervous system; COPD, chronic obstructive pulmonary disease; GI, gastrointestinal.

[†]Adjunct to a β-adrenoceptor antagonist.

[‡]Non-cancerous growth of nodules in the prostate, which can obstruct the urethra.

[§]Cancer of the adrenal gland.

(and dopamine) reuptake and is used to treat attention deficit hyperactivity disorder. Organophosphorus insecticides are irreversible acetylcholinesterase inhibitors and highly toxic, while physostigmine is a reversible enzyme inhibitor that can prevent mydriasis (dilation of the pupil).

Inside the presynaptic terminal the synthesis, storage, and release of the neurotransmitter from vesicles can also be disrupted. Increased release by storage disruption will, in the short term, have a stimulatory effect on the postsynaptic receptor, increasing synaptic levels of the neurotransmitter. However, if vesicular stores become depleted, responses will diminish until new transmitters can be synthesized. Furthermore, if over-stimulated, postsynaptic receptors can becomes less sensitive to the neurotransmitter release, reducing the response. Reserpine is an example of a chemical that can deplete stores of noradrenaline, and mebeverine is a related compound used in the treatment of irritable bowel syndrome (see Section 6.7)

Simply raising the levels of synaptic neurotransmitter would affect all synapses of that type and contribute to **non-specific side effects**. Box 6.2 considers a few examples where side effects occur owing to effects on the ANS.

BOX 6.2

The unwanted consequences of drug action

Specificity of action can be ensured by the use of agonists and antagonists selective for the receptors at the targeted terminals. Many drugs that are not specifically targeted at autonomic dysfunction affect the cholinergic or noradrenergic pathways and may have side effects that are a consequence of their autonomic actions. For example, the main action of some tricyclic antidepressants (for example amitriptyline) is inhibition of noradrenaline reuptake, increasing the level of synaptic noradrenaline. However, these drugs are also antagonists at muscarinic receptors. This activity in the periphery can produce parasympathetic side effects such as blurred vision, dry mouth, constipation, urinary retention, and tachycardia, which are particularly unpleasant for someone already struggling with depression.

Ganglion blockers

The autonomic ganglia are major control sites for the output to the ANS and can also be targeted by pharmacological agents. All preganglionic neurons are cholinergic, and at the level of the ganglia both pre- and postsynaptic receptors are nicotinic. Most acetylcholine enhancing agents will affect both these ganglionic receptors and those at the motor endplate of the somatic nervous system, and will therefore have both autonomic and neuromuscular consequences.

6.4 **Effect of the ANS on the cardiovascular system**

Physiology and pharmacology

Both the vasculature and the heart are innervated by the ANS, and the responses in these two organs are interrelated. Activation of the parasympathetic nervous system reduces the sympathetic activity by inhibiting the output of the *rostral ventrolateral medulla* in the brain (the centre for generating preganglionic sympathetic neuronal action potentials). Therefore it is generally true that the ANS mediates its effects on the cardiovascular system via the sympathetic nervous system, which is in turn regulated by the parasympathetic nervous system.

Blood vessels

The diameter of blood vessels is determined by the contractile tone of smooth muscle located in their walls. An increase in contractile tone causes a reduction in vascular diameter (vasoconstriction) and will increase the resistance to flow through that segment. A decrease in contractile tone causes an increase in vascular diameter (vasodilatation) and will decrease the resistance to flow through that segment. It is important to understand how resistance to flow is arranged in the circulation. Flow through organs is arranged in series, with the organs arranged in parallel, as shown in Figure 6.7. The implication of this arrangement is that an increase in resistance through one organ will cause a reduction

FIGURE 6.7 A schematic diagram of the flow of blood. Blood is oxygenated in the lungs and flows to the strong, thick-walled left side of the heart, which pumps it around the rest of body before it is returned to the thinner-walled right side of the heart to be pumped back to the lungs for re-oxygenation. Valves in the heart and blood vessels (yellow) prevent the blood from flowing in the wrong direction

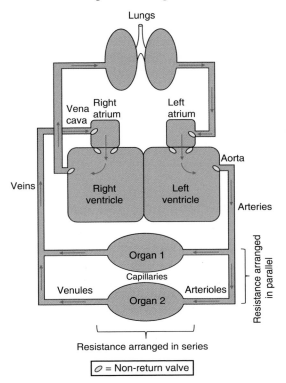

in flow through that organ and a redistribution of flow to the other organs.

Activation of the sympathetic nervous system affects the redistribution of cardiac output, the volume of blood pumped by the heart over 1 minute, by altering the diameter of blood vessels in different organs. It increases the flow of blood to the skeletal musculature, delivering oxygen and nutrients as well as removing metabolic waste products. Both β- and α-adrenoceptors are found on vascular smooth muscle. The $β_2$-adrenoceptor relaxes the smooth muscle, resulting in vasodilatation. The $α_1$-adrenoceptor is the predominant α-adrenoceptor but the $α_2$-adrenoceptor can also be found postsynaptically (see Table 6.2). Activation of either subtype of α-adrenoceptor causes contraction of smooth muscle, leading to vasoconstriction. The differential regulation of vascular responses comes about owing to the differential expression levels of $α_1$-, $α_2$- and $β_2$-adrenoceptors. A blood vessel containing mostly α-adrenoceptors will vasoconstrict in response to sympathetic activation whereas one containing mostly $β_2$-adrenoceptors will vasodilate.

There is comparatively little parasympathetic innervation of peripheral blood vessels as this is mainly confined to glandular and erectile tissue.

> **KEY POINT**
>
> The sympathetic nervous system redistributes blood flow to the skeletal musculature and away from the digestive organs. The parasympathetic nervous system has little effect on the vasculature and so does not directly cause a redistribution of blood flow.

> **SELF CHECK 6.4**
>
> In the fear response, if you have reduced production of saliva and increased blood activity of the muscles, which receptor predominates in the blood vessels of each organ?

Heart

The key parameter of cardiac function is **cardiac output**, defined as **stroke volume** multiplied by heart rate. Put simply, cardiac output is the rate at which blood volume is ejected from the heart. It is a common misconception to think that the main function of the heart is to generate pressure when, in fact, *pressure generation is a product of the need to generate cardiac output*. We will now examine how the ANS affects cardiac function in terms of heart rate and force of contraction.

- *Heart rate*. Activation of the sympathetic nervous system leads to a profound increase in heart rate. This response is mediated by the β_1-adrenoceptor. Both noradrenaline (released from sympathetic nerve **varicosities**) and circulating adrenaline (released from the adrenal medulla) can activate β_1-adrenoceptors located in the *sinoatrial (SA) node* of the heart. The SA node is a specialized structure located at the junction of the right **atrium** and the **superior vena cava**. It is responsible for the spontaneous generation of action potentials, and determines heart rate.

  In the heart, activation of the parasympathetic activation elicits a profound slowing of the heart rate. This is achieved by acetylcholine, released from postganglionic parasympathetic nerves, stimulating muscarinic M_2 receptors located in the SA node.

- *Force of contraction*: The force of cardiac contraction is increased in response to activation of the sympathetic nervous system. This response is mediated mostly by β_1-adrenoceptors located on the **myocytes** of the ventricular muscle. Through a complex molecular pathway, the amount of calcium (Ca^{2+}) released into the **sarcoplasm** of a **cardiomyocyte** during a contraction cycle is increased, as explained in Integration Box 6.1. The amount of free Ca^{2+} in the sarcoplasm governs the magnitude of the contraction. An increase in the force of cardiac contraction increases the pressure difference between the left **ventricle** and the venous system. Stimulation of muscarinic M_2 receptors can also affect the force of contraction in the ventricles, although this is a relatively minor response, being more effective in the atria. The parasympathetic nervous system decreases the force of contraction as the level of free Ca^{2+} in the sarcoplasm during a contraction cycle decreases.

  Integration Box 6.1 looks into the effect of Ca^{2+} on the regulation of the force of contraction.

SELF CHECK 6.5

What are the consequences of increased force of contraction on blood flow through tissue?

INTEGRATION BOX 6.1

Force of contraction is regulated by calcium ions

Calcium is considered the universal second messenger owing to its role in signal transduction in the whole body. In the heart the force of contraction is closely related to the level of unbound Ca^{2+} in the sarcoplasm, whereby increases in free Ca^{2+} increase the force of contraction. There are two sources for this sarcoplasmic Ca^{2+}, extracellular and intracellular, which are linked. Firstly, extracellular Ca^{2+} enters the cardiomyocyte through voltage-sensitive **L-type Ca^{2+} channels**. As these channels are voltage sensitive they only open during **depolarization**. Once Ca^{2+} enters the cardiomyocyte it triggers the release of intracellular stores of Ca^{2+}. The sum of these sources accounts for the Ca^{2+} that regulates contraction. Alteration in either of these sources will affect the amount of free Ca^{2+} in the sarcoplasm and hence contraction.

 Section 5.3 'How do nerve cells communicate with each other?', within Chapter 5 of this book, introduces calcium as a second messenger. Calcium homeostasis is discussed in Section 8.3 'Major homeostatic mechanisms in the human body', within Chapter 8 of this book.

In the cardiomyocytes the parasympathetic and sympathetic nervous systems have contrasting effects on the sarcoplasmic concentration of Ca^{2+}, resulting in opposing outcomes on the force of cardiac contraction.

Sympathetic nervous activity results in the elevation of Ca^{2+} levels and hence an increase in the force of contraction, whereas activity of the parasympathetic nervous system will cause a decrease in Ca^{2+} levels and a consequential reduction in the force of contraction. Many of the drugs used for the treatment of cardiovascular disorders, for example the calcium channel-blocking drugs verapamil and nifedipine, affect these interlinked pathways in some way.

Therapeutics

Currently, the parasympathetic nervous system provides little opportunity for pharmacological interventions targeting cardiovascular disorders. The reason for this is the lack of pharmacological agents with sufficient nicotinic or muscarinic receptor subtype selectivity to be useful as therapeutic agents. The neurotransmitter acetylcholine is common to the neuromuscular junction as well as preganglionic autonomic neurotransmission.

SELF CHECK 6.6

Why is the sympathetic nervous system a more attractive therapeutic target for cardiovascular disorders than the parasympathetic nervous system?

Non-selective sympathomimetic agonists, such as adrenaline and noradrenaline, are adrenoceptor agonists that activate both α- and β-adrenoceptors with a similar potency. They are normally used only in emergency situations, as the lack of pharmacological selectivity can lead to significant side effects with prolonged use. Nonetheless, these endogenous molecules can be useful clinically as short-acting agents because they are quickly metabolized. Adrenaline can be used to try and restore cardiac rhythm, although there is little clinical evidence that it has any significant benefit in this setting. Noradrenaline has a greater clinical effect on α-adrenoceptors than adrenaline and is used to increase blood pressure by inducing peripheral vasoconstriction leading to an increase in vascular resistance.

β-Adrenoceptor agonists such as dobutamine selectively activate β-adrenoceptors. Generally they are used to increase blood pressure in conditions such as clinical shock (an acute pathological fall in blood pressure). The key β-adrenoceptor subtype mediating their clinically useful activity is the $β_1$-adrenoceptor, activation of which increases cardiac workload. The use of agents that increase cardiac workload needs to be carefully considered in patients with myocardial infarction or any other type of ischaemic heart disease, as they will increase metabolic demand leading to greater tissue damage.

β-Adrenoceptor antagonists such as atenolol, carvedilol, metoprolol, and labetalol are mainly used to treat hypertension. Again, the important receptor subtype is the $β_1$-adrenoceptor. Although antagonism of cardiac $β_1$-adrenoceptors can contribute to a reduction in pressure by reducing heart rate and force of contraction, the beneficial long-term effects in reducing blood pressure are mediated by the kidney (see Section 6.9). $β_1$-Adrenoceptor antagonists can also be used in the treatment of heart failure, although angiotensin-converting enzyme (ACE) inhibitors are the clinical drugs of choice.

There are a number of potential adverse effects associated with the use of $β_1$-adrenoceptor antagonists but the most important contraindication is for patients with asthma. Currently available $β_1$-adrenoceptor antagonists have only a relatively modest selectivity for $β_1$- over $β_2$-adrenoceptors. This has particular significance for asthmatics because the $β_2$-adrenoceptor activation in the lung is of great importance in dilating the airway to provide relief from an asthma attack.

$α_1$-Adrenoceptor agonists such as phenylephrine are now mainly used in over-the-counter palliative medications for colds and influenza. The mechanism of action is to cause vasoconstriction that reduces glandular secretions, thereby reducing congestion in the nasal passages and sinuses. These should only be used for short periods of time (no more than once every 3 h for no longer than 3 weeks), as by reducing blood flow through tissue there can be a failure to meet metabolic demand, resulting in tissue damage.

$α_2$-Adrenoceptor agonists such as clonidine can be used to lower blood pressure. This is somewhat surprising as $α_2$-adrenoceptors are found postsynaptically in blood vessels, where they mediate vasoconstriction. The site of action is now thought to be in the brain,

where α_2-adrenoceptor stimulation causes a reduction in sympathetic nervous outflow.

α_1-*Adrenoceptor antagonists* such as prazosin lower blood pressure by causing vasodilatation, resulting in reduced peripheral vascular resistance. Postural hypotension (a marked fall in blood pressure upon standing upright) is a significant adverse effect that can cause syncope (loss of consciousness).

6.5 Effect of the ANS on the respiratory system

Physiology and pharmacology

Little is known of how the ANS regulates pulmonary function. Autonomic innervation of the lungs is mainly parasympathetic, although there is still significant sympathetic input via circulating adrenaline. The predominant receptor mediating sympathetic responses is the β_2-adrenoceptor located on bronchial smooth muscle. Activation of β_2-adrenoceptors results in activation of the stimulatory G-protein G_s. This causes relaxation of the bronchial smooth muscle (*bronchodilation*) causing airway diameter to increase and resistance to airflow to be reduced. Physiologically, this is useful under conditions of sympathetic nervous system activation (such as exercise) as it facilitates an increase in air movement to and from the lungs.

Bronchial smooth muscle receives significant parasympathetic nervous input. Muscarinic M_3 receptors are found on bronchial smooth muscle and mediate contraction causing a narrowing of the airway (*bronchoconstriction*) via activation of another type of stimulatory G-protein, G_q. There are also M_2 receptors located on the presynaptic terminals of postganglionic neurons, which act to inhibit the neuronal release of acetylcholine via activation of the inhibitory G-protein, G_i.

Therapeutics

Therapeutically, the β_2 receptor is very important in the treatment of asthma. Different agonists selective for β_2-adrenoceptors, of which the most well known is salbutamol, are an important class of therapeutics used to relieve the effects of an asthma attack. There are other therapeutically useful β_2-adrenoceptor agonists that differ from salbutamol only in their duration of action. Long-acting β-adrenoceptor agonists, of which formoterol is an example, are used to provide longer-lasting bronchodilator coverage when attacks are more frequent. This is further considered in Case Study 6.1.

In the parasympathetic system the aim is to antagonize muscarinic M_3-mediated bronchoconstriction. Currently available drugs, such as tiotropium, have little selectivity between muscarinic receptor subtypes and so their effectiveness is likely to be limited by antagonism of presynaptic M_2 receptors, which will cause an increase in acetylcholine outflow. Clinically, drugs such as tiotropium are not widely used in the treatment of asthma except when the symptoms are not adequately controlled by conventional treatment. There is more widespread use of inhaled muscarinic antagonists in chronic obstructive pulmonary disease, where β_2-adrenoceptor agonists are less effective.

CASE STUDY 6.1

Dilip has hypertension (high blood pressure) which has been well controlled using the β-adrenoceptor blocker atenolol. He went to the GP because he had been experiencing shortness of breath when walking a distance. The doctor sent him to the hospital to have cardiac and respiration investigations done and cardiac results showed that his heart was the normal size and there was nothing wrong with his lungs. The GP said he should stop taking atenolol and prescribed captopril and a formoterol inhaler.

REFLECTION QUESTIONS

1. What is the most likely diagnosis and treatment for Dilip's condition?

2. Why does Dilip have to stop taking the atenolol and change to captopril?

3. Why might formoterol be prescribed instead of salbutamol for asthma?

4. Why might switching to the ACE inhibitor to control hypertension actually be more beneficial in the long run?

Answers

1 The shortness of breath is not due to hypertrophy or enlargement of the heart, which is a sign of heart failure, and the respiratory tests showed nothing. The fact that Dilip is experiencing shortness of breath when walking suggests it is probably exercise-induced asthma that is causing the symptoms.

2 The β₁ receptor antagonist atenolol works by blocking β receptors in the heart, acting through the sympathetic nervous system, competing with the noradrenaline released by the synapse, resulting in reduced heart rate and force of contraction. Formoterol is a long acting β₂ agonist, which will act to stimulate the sympathetic nervous system to relax the bronchioles, increasing airflow into the lungs.

Both atenolol and formoterol would therefore compete at the same receptors in the same organs and thus would not be effective. Captopril works by a different mechanism, inhibiting angiotensin-converting enzyme (or ACE). Formoterol is a long acting β₂ agonist, which will act to stimulate the sympathetic nervous system to relax the bronchioles, increasing airflow into the lungs.

The primary action of the ACE inhibitor captopril is in the direct blockade of the conversion of angiotensin I to angiotensin II. Angiotensin II stimulates sympathetic activity directly and increases adrenaline release at the adrenal gland. The ACE inhibitor will reduce the amount of angiotensin II reducing cardiac load by numerous mechanisms including reduced peripheral vasoconstriction, reduced blood volume, and reduced sympathetic output to the heart. All of these mechanisms are beneficial in reducing blood pressure.

3 Formoterol is a long acting β₂ agonist of up to 12 hours as opposed to the far shorter acting salbutamol. This would allow relief from the shortness of breath throughout the day when doing any kind of activity, which is more appropriate for this form of asthma.

4 Long-term ACE inhibitors are a preferred drug as they are not associated with increased risk of type II diabetes and have been shown to have a long-term effect on reducing the risk of congestive heart failure.

6.6 **Effect of the ANS on the oculomotor system**

Physiology and pharmacology

The direction our eyes move is voluntarily controlled by three cranial nerves that innervate six muscles around the eyeball itself. However, clear vision requires focusing of the lens and adjustments for the amount of light entering the eye. These intrinsic eye muscles are innervated by the ANS.

The *ciliary muscle*, shown in Figure 6.8(A), controls the *pliable lens* of the eye. In a process called *accommodation*, the ciliary muscle changes the shape of the lens to bring objects into focus regardless of their distance. It is innervated by M_3 receptors on parasympathetic fibres from the ciliary ganglion.

Limiting the amount of light that reaches the back of the eye is critical to avoid damaging the retina. The size of the pupil, the dark centre of the eye in the middle of the coloured iris, controls the amount

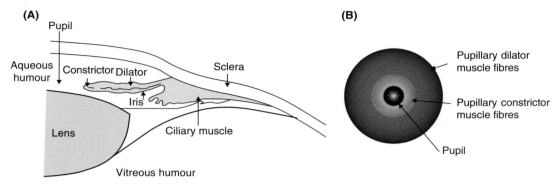

FIGURE 6.8   The anatomy of the eye. (A) The ciliary muscle controls the shape of the eye and is therefore useful in adjusting the focus of the lens. (B) The constrictor and dilator parts of the iris, which contain circular (constrictor) and radial (dilator) fibres that are controlled by the opposing divisions of the autonomic nervous system

of light entering the eye (shown in Figure 6.8(B)). The iris receives sympathetic and parasympathetic innervation to the radial pupillary dilator and circular pupillary constrictor muscles, respectively. Both are tonically active during the waking state, adjusting in response to emotional stimuli and the environment. Darkness increases sympathetic tone, contracting the dilator muscle fibres and causing pupillary dilation (causing those big romantic eyes over a candlelit dinner). Conversely, light stimulates the parasympathetic nervous system to contract the circular muscle, constricting the pupil.

Assessing baseline pupillary size can give some superficial indication as to the cause of illness if failure of these systems is suspected: a small pupil indicates sympathetic dysfunction while a large pupil is indicative of parasympathetic dysfunction. The sympathomimetics amphetamine and cocaine will increase sympathetic tone and give the characteristic big pupils associated with these drugs of abuse.

Therapeutics

Clinically, the advantage of administering drugs to the eye is that local effects can be produced with eye drops, which avoid the generalized autonomic and central effects produced by systemic administration. For ophthalmic use, non-selective muscarinic antagonists such as tropicamide and cyclopentolate are used to dilate the pupils.

Glaucoma is associated with raised intraocular pressure, which may damage the eye if left untreated. Constriction of the pupil allows fluid to drain from the aqueous humour within the eye, thus reducing the pressure. Muscarinic agonists such as pilocarpine will activate the M_3 receptors on the parasympathetic pupillary constrictor muscle, leading to constriction of the pupil. Alternatively, β-adrenergic antagonists can be applied to relax the ciliary muscle, which may also facilitate the loss of fluid from the eye and lower the intraocular pressure.

6.7   Effect of the ANS on the digestive system

Physiology and pharmacology

The digestive system includes the gastrointestinal (GI) tract (stomach and intestines), but digestion actually starts with salivary enzymes in the mouth and can only take place with help from the liver, gall bladder,

and pancreas (dealt with in Section 6.10). Saliva is produced during mastication (chewing) and contains salivary α-amylase and mucins, which aid in the initial breakdown of food and ease its passage down the gullet. The salivary glands are predominantly controlled by the parasympathetic nervous system. Parasympathetic stimulation increases secretion of saliva and blood flow to the salivary glands. The sympathetic nervous system can also stimulate the secretion of saliva and increase the amount of amylase, but vasoconstriction reduces blood flow to the salivary glands and ultimately reduces saliva production (hence the dry mouth when nervous). Food is swallowed and although this process is voluntary, autonomic reflex processes continue the movement of food by peristaltic contractions down the gullet.

The GI tract is made up of smooth muscle that is highly extrinsically innervated by both autonomic systems. Through a combination of enzymatic secretions and mechanical breakdown of food by rhythmical contractions, the digestive process continues in the stomach and down into the intestines. The sympathetic nervous system reduces motility, slowing digestion (via α_1-, α_2-, and β_2-adrenoceptors), whilst the parasympathetic 'rest and digest' innervation increases gut motility and produces gastric acid and enzymatic secretions (via M_1/M_3 receptors), thus encouraging digestion.

Chemo- and mechanoreceptors in the GI tract are stimulated by the presence of food or distension of the gut wall and send this information locally and back to the CNS via afferent nerves. This induces a reflex response by the parasympathetic nervous system to produce gastric secretions. The sphincters present in the GI tract to control the flow of contents are contracted by the sympathetic nervous system and relaxed by the parasympathetic nervous system.

Therapeutics

Parasympatholytic muscarinic antagonists such as hyoscine and mebeverine are used to reduce the contractile activity of the GI tract, especially if spasms produce painful cramps or nausea and vomiting. This is useful in the treatment of motion sickness and irritable bowel syndrome, as mentioned in Case Study 4.1. Mebeverine is a derivative of reserpine, a compound that depletes neuronal stores of catecholamines thus causing relaxation via stimulation of the sympathetic nervous system. Muscarinic side effects include dizziness, dry mouth, blurred vision, and sedation. Small doses of *atropine*, a non-selective muscarinic antagonist, have been used to treat diarrhoea but because of its general muscarinic effects it is rarely used.

6.8 Effect of the ANS on the hepatic system

Physiology and pharmacology

Located between the gut and circulatory system, the four lobes of the liver carry out a range of functions. It is the site of bile and protein synthesis, lipid and lipid-soluble vitamin storage and metabolism, and manages the excretion of various endogenous and exogenous toxic products. Many of the functions of the liver are beyond the scope of this chapter as they are not regulated by the ANS but *glycogenolysis*, the breakdown of the glucose storage product glycogen, is regulated by the sympathetic nervous system. In addition *gluconeogenesis*, the synthesis of glucose from non-carbohydrate substances, can also help control blood glucose levels and is under the same control.

 Glucose metabolism is described further in Section 3.3 'Metabolic pathways and abnormal metabolism', within Chapter 3 of this book.

The pancreas produces insulin, which promotes the uptake of glucose by the liver. This glucose is transformed into glycogen and stored by the liver. When insulin levels decline the glycogen is broken down to release glucose back into the bloodstream.

Glycogenolysis is under autonomic control and the blood supply to the liver is innervated by neurons of the sympathetic and parasympathetic pathways. In addition, sympathetic projections innervate deep into the structure, directly onto hepatocytes, which are likely to be the site of direct control of the ANS over glucose production. Sympathetic activity stimulates glycogenolysis and gluconeogenesis while the para-sympathetic vagus nerve accelerates the synthesis of glycogen, reducing blood glucose. The vagus nerve contains sensory afferent nerves that sense blood glucose levels and inhibit its own parasympathetic activity, but the function and consequence of this action is unclear. It also triggers the release of *hepatic insulin-sensitizing substance* affecting the control of skeletal muscle glucose uptake, which is elevated in exercise or stress.

The green sac of the gall bladder, coloured by the bile it contains, is located next to the liver. Bile made by the liver is stored there ready to be released once a meal begins. At this point the parasympathetic ner-vous system (M_3 muscarinic receptor mediated) contracts the smooth muscle of the gall bladder, forcing the bile into the duodenum where it emulsifies fats for digestion. The β-adrenoceptors of the sympathetic nervous system oppose this action, relaxing the smooth muscle and inhibiting bile secretion.

Therapeutics

Neither the liver nor the gall bladder are direct targets of autonomic pharmacotherapy. Their regulation serves to increase the amount of glucose available to the muscles and circulation, when required, while reducing the production of digestive secretions.

6.9 Effect of the ANS on the renal and urinary system

Physiology and pharmacology

Kidney

Our varied dietary intake and the demands of our environment (temperature, humidity etc.) all create challenges in the regulation of the volume, electrolyte content, and pH balance of extracellular fluid. The kidneys play an essential role in regulating each of these, reabsorbing useful elements (the majority of water and sodium, potassium, and chloride ions), and filtering out the water-soluble waste products. Several hundred litres of extracellular fluid each day pass through structures within the **nephrons** of the kidneys called the *juxtaglomerular apparatus*, but adequate filtration at this site relies upon the maintenance of pressure. The kidneys secrete an enzyme called *renin*, which cleaves a precursor protein, angiotensinogen, into angiotensin I, which is further cleaved into angiotensin II by ACE. This highly active peptide acts on the adrenal glands to release steroid hormones, the consequence of which is increased blood volume and vasoconstriction. The resulting increase in blood pressure then allows the kidneys to function appropriately. Renin release is triggered by reduced blood pressure and the activation of the sympathetic innervation to the kidney.

Bladder

Although we have stated that the ANS controls functions without conscious thought, in some organs there is interplay with the voluntary system. The urinary system illustrates just that, as we know we have voluntary control over urination. As the bladder fills with urine, mechanoreceptors activate the sensory neurons at the sacral level of the spinal cord (S2 to S4). In a process known as the *micturition reflex*, the parasympathetic innervation to the *detrusor* smooth muscle of the bladder initiates contraction. If urination at that moment is not desired, this reflex is overridden by higher centre control in the brain. Contraction of the skeletal muscle fibres of the external sphincter muscle maintained by the somatic motor system prevents urination. In addition, the preganglionic sympathetic neurons extend from lumbar regions (L1 and L2) of the spinal cord and noradrenergic postganglionic neurons also innervate the detrusor muscle and internal urethral sphincter. These work in opposition to the parasympathetic innervation, maintaining contraction of the sphincter and relaxation of the bladder. Temporary inactivity of the somatic motor innervation coordinated with parasympathetic mediation of bladder contraction and relaxation of the internal sphincter allows voluntary urination to occur when desired.

The presence of the bladder reflex is well illustrated in patients with spinal cord injury at the sacral level, and in small infants. They are unable to control their urination voluntarily and do not continuously release urine, but rather the reflexive bladder contracts automatically when it becomes sufficiently distended to trigger voiding at intervals.

Therapeutics

Kidney

Sympathetic activation of β_1-adrenoceptors in the kidney causes the release of renin, which leads to the production of angiotensin II. Angiotensin II is an important enzyme in blood pressure regulation as it is not only a vasoconstrictor but also leads to water retention. Inhibition of renin production and consequent reduction of angiotensin II levels results in reduction of blood volume, which is the most effective way to reduce blood pressure over the long term. This

is one of the reasons that *ACE inhibitors*, which block the key step in angiotensin II production, are superseding the use of β_1-adrenoceptor antagonists in the treatment of hypertension.

Bladder

The sympathetic contraction of the internal sphincter and relaxation of the bladder are mediated through the noradrenergic α_1- and β_2- adrenoceptors, respectively, while the M_3 cholinergic receptor receives the parasympathetic input. Benign prostatic hyperplasia is an enlargement of the prostate generated by noncancerous overgrowth of epithelial cells. This growth increases pressure on the urethra, which leads to problems with frequency and urge to urinate. Long term, this is alleviated by surgery, but α_1-adrenoceptor antagonists such as tamsulosin can relieve some of the symptoms. Bed wetting (or nocturnal enuresis) in older children may warrant medical intervention; tricyclic antidepressants, which inhibit noradrenaline reuptake and have muscarinic antagonist properties, can be used to increase sympathetic tone and reduce parasympathetic tone and thereby reduce involuntary urination. However, although in principle manipulating autonomic function should assist in urinary dysfunction, balancing incontinence with urinary retention is very difficult.

6.10 **Effect of the ANS on the endocrine systems**

Physiology and pharmacology

The secretory processes of human physiology involve organs that release substances into the external environment (*exocrine*) or into the bloodstream (*endocrine*). A series of glands synthesize and secrete hormones, signalling target tissues in a far slower and more prolonged response than the rapidly acting nervous system. In general terms the sympathetic nervous system inhibits secretions by vasoconstriction whereas the parasympathetic nervous system stimulates secretion. The nasal, lacrimal, and salivary glands, as well as those of the GI system and the

pancreas, are all secretory organs controlled in this fashion; however, the adrenal medulla is a significant exception to this rule.

Pancreas

The pancreas is a major exocrine organ, secreting digestive enzymes that break down dietary carbohydrates, fats, and proteins. However, the pancreas is also considered an endocrine gland, as it secretes insulin and glucagon into the bloodstream, which have opposing actions for the essential minute-by-minute control of blood glucose levels. The *islets of Langerhans* are the location of the *β-cells*, which synthesize and secrete insulin, while glucagon is secreted from *α-cells*.

The sympathetic nervous system decreases insulin secretion, elevating the amount of available glucose during increased activity of organs that have high energy requirements. This is opposed by the parasympathetic nervous system increasing insulin secretion.

Adrenal gland

The adrenal gland is another endocrine organ that has a particular relationship with the ANS. The adrenal gland consists of the outer *adrenal cortex* and the inner *adrenal medulla*. It is the only organ to receive direct autonomic innervation from the spinal cord without passing through ganglia and is solely innervated by the sympathetic nervous system (as shown in Figure 6.3). The sympathetic neuron that innervates it is cholinergic, as are all preganglionic neurons, and the response is mediated by postsynaptic nicotinic receptors. These differ from the muscarinic receptors of the parasympathetic nervous system target organs in that they are ion channels, which open in response to ligand binding. The release of acetylcholine causes the influx of ions which depolarize the cell membrane and trigger a rapid response in the adrenal medulla to secrete noradrenaline (20%) and adrenaline (80%). These catecholamines, key to the physiological events of the 'fight or flight' response, are then circulated through the body. In this context both molecules are acting as hormones because they may target distant organs indirectly to promote the actions of the sympathetic nervous system: increased cardiac output, bronchodilation increasing oxygen flow into the lungs, and reduced gut motility conserving and redirect-

ing energy to where it can be more useful. The rapid metabolism of the catecholamines by circulating catechol O-methyltransferase means that the responses are relatively short lived and well regulated.

Therapeutics

Pancreas

The secretory activity of the pancreas is largely controlled by the action of hormones in the circulating blood. However, α_2-adrenoceptor sympathetic activity decreases insulin production and increases glucagon release, providing more glucose to be used by the energetic demands of the flight response. β_2-Adrenoceptor activity increases β-cell secretions. Conversely, the parasympathetic innervation controlled by the M_3 receptors increases secretion of both insulin and glucagon. None of these, however, are therapeutic targets.

Adrenal gland

Cancerous overgrowth of the adrenal gland results in overproduction of adrenaline and noradrenaline. This hyperfunctioning of the adrenal gland results in sporadic symptoms of excessive sympathetic activity, including hypertension and headaches, sweating and palpitations, as well as causing CNS effects such as anxiety. There can also be converse down-regulation of noradrenergic receptors as a result of this adrenergic overload, leading to postural hypotension. This is typically treated with surgery to remove the tumour but the administration of α- and β-adrenergic receptor antagonists will control the symptoms and is a critical step prior to surgery to avoid a sudden release of catecholamines when the tumour is disturbed.

6.11 Effect of the ANS on the reproductive system

Physiology and pharmacology

Penile and vaginal innervation by the ANS is an unusual example of the two autonomic systems working together rather than in opposition, and has similar effects in males and females. Parasympathetic activity

causes erection of the sexual organs, both the penile and clitoral tissue, while the sympathetic nervous system controls ejaculation in the male and vaginal contraction and mucus secretion in the female. The uterus is controlled by the sympathetic nervous system, both contraction and relaxation mediated through different adrenergic receptors.

Therapeutics

There are no autonomic drugs targeted to treat sexual dysfunction but antidepressant and antihypertensive agents can both have the side effect of impairing sexual function through their muscarinic antagonist activity. α_1-Adrenoceptors mediate contraction whereas β_2-adrenoceptors relax the uterus. Therefore spontaneous or hormone-induced contractions of the pregnant uterus, which could lead to premature birth of the baby, can be inhibited with the use of β_2-adrenoceptor agonists such as ritodrine or salbutamol. Delaying delivery of a premature baby can allow the administration of steroids to mature the lungs in anticipation of the birth.

6.12 Effect of the ANS on the integumentary system

Physiology and pharmacology

The integumentary system is largely composed of the skin but also includes the hair, nails, and exocrine glands located within or on the skin. It has an array of functions to protect, regulate, and excrete as well as to contain sensory receptors or corpuscles to detect temperature, pain, and pressure.

Sweat glands in the skin are an example of ducted exocrine glands. They are controlled by the sympathetic nervous system (no action from the parasympathetic branch) but instead of the postganglionic neuron being noradrenergic it is cholinergic. The secreted acetylcholine activates the M_3 muscarinic receptors on the sweat glands, stimulating the sweat response. Sweat glands are activated when the external temperature increases, and in times of stress, resulting in increasing sweating which cools the body. There is also some contribution from the adrenergic α_1-adrenoceptors, which are also responsible for activity on the *arrector pili muscles*, attached to the hair follicles in the skin. When the external temperature decreases, or adrenaline surges through fear, the arrector pili muscles contract, the hairs become erect and we get 'goose bumps'.

 Temperature regulation is discussed in more detail in Section 8.3 'Major homeostatic mechanisms in the human body', within Chapter 8 of this book.

Therapeutics

Clinically, there are few uses for autonomic agents acting on the integumentary system but muscarinic agonists would increase sweating through the M_3 receptor. Poisoning from the muscarinic receptor antagonist atropine, which can occur in children who have eaten deadly nightshade berries, reduces sweating, which contributes to a potentially dangerous rise in body temperature.

6.13 Non-adrenergic non-cholinergic control of the ANS

The term *non-adrenergic non-cholinergic control* (NANC) refers to transmission in the nervous system by transmitters other than noradrenaline and acetylcholine. It accounts for the many effects in the ANS that cannot be attributed to either of the recognized pathways. NANC neurotransmitters vary depending on the location and target organ but include both peptide and non-peptide entities. Co-transmitters can either reside within the adrenergic or cholinergic presynaptic vesicles; for example, noradrenergic fibres may contain adenosine-5′-triphosphate (ATP) or gonadotrophin-releasing hormone or, alternatively, like neuropeptide Y (NPY), may be contained within separate vesicles. These co-transmitters are released at the same time as the major neurotransmitter, binding to their own receptors on the presynaptic (ATP and NPY) or postsynaptic (ATP) membrane.

In some actions of the sympathetic nervous system these co-transmitters can dominate the response or feedback to inhibit further noradrenaline release.

The additional control and selectivity that this process affords may be mediated by the fact that the co-transmitter may vary from organ to organ, may be present in the synaptic cleft for a different length of time to that of the major neurotransmitter, may be preferentially released at certain frequencies of stimulation, or be differentially regulated by presynaptic feedback receptors. Co-transmitters can be present in either pre- or postganglionic terminals—for example, the peptide substance P is an acetylcholine co-transmitter at sympathetic ganglia. The enteric nervous system utilizes a range of co-transmitters; up to 20 different entities have been identified in terminals including γ-aminobutyric acid (GABA) and serotonin (5-hydroxytryptamine, 5-HT). Our understanding of these NANC processes is developing, but their involvement in current therapeutics is limited.

CHAPTER SUMMARY

➤ The two divisions of the autonomic nervous system (ANS) are constantly active, producing the necessary level of tone for our bodies to function. When the body is required to change in response to an event or stimulus the sympathetic drive increases, allowing the body to respond appropriately.

➤ The sympathetic nervous system can enhance its own function by releasing adrenaline from the adrenal medulla, which acts in the same way as noradrenaline but reaches the organs by circulating in the blood as opposed to being released at the synapse.

➤ The ANS is driven by two different anatomical pathways, which use two different neurotransmitter systems: noradrenergic (releasing noradrenaline) and cholinergic (releasing acetylcholine). This means that each system can be targeted pharmacologically, and classes of drugs have been developed and used successfully in the treatment of disease. However, because these two systems affect so many areas of the body there are significant unwanted effects with many of these drugs.

➤ Although these two parts of the ANS act to a large extent through reflex responses, there is overall control mediated by the hypothalamus, with influence from the cerebrum, which is reflected in the physiological response to strong emotions.

➤ The distribution of different receptor subtypes and the type of G-protein they work through means that the sympathetic nervous system can affect parts of the body differently, depending on the receptors that are expressed, and this is how blood vessels in different areas are regulated.

➤ Drugs can directly target receptors or can function through enhancing the activity of the synapse or neurotransmitter.

➤ The cardiac system is heavily controlled by the autonomic system and can be influenced by a range of drugs, depending on the medical condition that requires treatment.

➤ To add complication to this system, other neurotransmitters can also be released alongside noradrenaline and acetylcholine, and act on different organs to facilitate or counteract the effect.

FURTHER READING

NICE clinical guideline CG127. *Hypertension: Clinical Management of Primary Hypertension in Adults.* National Institute for Health and Clinical Excellence, 2012. <http://publications.nice.org.uk/hypertension-cg127>.

These guidelines describe the reasons for the choice of drug in treating hypertension, as would be considered in Case Study 6.1.

Randall, M.D. and Neil, K.E. *Disease Management: A Guide to Clinical Pharmacology.* 2nd edn. Pharmaceutical Press, 2009.

This book is of interest because it covers in greater detail the therapeutic management of the autonomic diseases described in this chapter.

7

Communication systems in the body: autocoids and hormones

PETER N. C. ELLIOTT

Human beings are very complex organisms made up of many billions of cells formed into tissues and organs. To function effectively and efficiently it is imperative that the activities of cells are coordinated. To achieve this coordination there must be communication between cells, tissues, and organs throughout the body. There are basically four levels of communication: via nerves, autocoids, and hormones as well as direct communication between adjacent cells. Cells can communicate directly with each other through surface interaction. Cell membranes have a large variety of surface structures which are capable of interaction with adjacent cells. This interaction between cells relies on the very local exchange of particular chemicals known as cytokines, which induce the recipient cell to react in a particular way. You may have already read about neural communication in Chapters 5 and 6 of this book, while the focus in this chapter will be on communication within the body via autocoids (also known as *local* hormones) and hormones (considered *true* hormones). We will explore how these endogenous molecules are involved in the development of the symptoms of pathologies such as hay fever, and consider drugs that are used to alleviate these symptoms.

Learning objectives

Having read this chapter you are expected to be able to:

➤ Understand the general role of autocoids and true hormones within the body.

➤ Describe the role of autocoids as a means of communication within the body.

➤ Appreciate the role of histamine in health and disease.

➤ Understand the role of serotonin in the body.

➤ Show an understanding of the synthesis and properties of kinins and eicosanoids.

➤ Understand the nature and role of hormones.

7.1 Autocoids

Autocoids (local hormones) are chemicals released by cells that will affect only those cells in close proximity to the source of their release. There is no mechanism for distribution, as is the case for true hormones (considered in Section 7.6), which are distributed by the blood throughout the body. Local hormones are disseminated simply by diffusion. If nerves give fine-level control using an instant, short-acting, targeted messaging system, and hormones give rise to longer-lasting, 'scene-setting' control, local hormones give an intermediate level of control.

Autocoids (derived from the Greek meaning *self-medicating*) include a range of different molecules as shown in Table 7.1. We will look into each of these in more detail in this chapter.

The establishment of an autocoid role generally follows the discovery that a molecule found in the body can induce particular responses when applied to the tissues. Of course, it is important to appreciate that the mere demonstration that a molecule found in the body possesses a particular biological effect does not prove that it is involved in any control mechanisms. Many such substances can produce all sorts of pharmacological effects in the body when administered but some of these effects may only be seen when large quantities (greater than those occurring in normal tissues) are applied to the body. Although a substance exhibits a pharmacological property, it may be a false premise to assume that it has a physiological role.

> **KEY POINT**
>
> Autocoids are chemicals that are released by cells and which affect the activities of other cells in the locality. Their distribution is by diffusion and they are usually inactivated within seconds.

TABLE 7.1   A range of different autocoids and their actions

Autocoid	Nature	Properties
Histamine	Amine derived from the amino acid histidine	Vasodilatation
		Bronchoconstriction
		Smooth muscle contraction
Serotonin	Also known as 5-hydroxytryptamine (5-HT), an amine derived from the amino acid tryptophan	Vasoactive
		Neurotransmitter
		Gastrointestinal smooth muscle stimulant
Bradykinin	Peptide containing nine amino acids	Vasodilatation
Prostaglandins	Unsaturated fatty acids containing 20 carbon atoms and with an internal cyclic structure	Wide variety of effects
Leukotrienes	Straight-chain unsaturated fatty acid containing 20 carbon atoms (some also have an amino acid substituent)	Vasoactive
		Bronchoconstriction
		Chemotactic
Thromboxane	Unsaturated fatty acid containing 20 carbon atoms and having an internal cyclic structure	Platelet aggregation

7.2 Histamine

Perhaps the most obvious example of a locally acting hormone is *histamine*. Histamine is an example of a biologically active amine and is found in many mammalian tissues as well as in plants and some insect venoms. Interest in histamine first arose when physiologists noticed that it could induce a reaction in mammals that was very similar to anaphylactic shock. Subsequently histamine was shown to be released during anaphylaxis.

Histamine is synthesized in the body by the decarboxylation of the essential amino acid *histidine*, as shown in Figure 7.1. It can also be ingested as part of normal food intake, as described in Box 7.1.

Following its release, histamine is rapidly metabolized to inactive products. It is first methylated by histamine *N*-methyltransferase to give methylhistamine, which is then oxidized by diamine oxidase resulting in methylimidazoleacetic acid. Histamine may also be directly oxidized by diamine oxidase to give imidazoleacetic acid. The inactive metabolites of histamine are then excreted in the urine. The activity of histamine, once released, is therefore of limited duration.

Histamine is always present in epithelial tissues, where it is found stored in mast cells. It is also found in basophils in the blood. Within both of these cells histamine is stored in *cytoplasmic granules* in association with heparin. Release may be achieved by exocytosis of the granules or by disintegration of the cell. Release from mast cells is triggered by exposure to an allergen to which the individual is sensitive. Once released, histamine interacts with *histamine receptors*, which are a specific type of G-protein-coupled receptor, present on the surfaces of various cells.

FIGURE 7.1 **Histamine synthesis occurs via the decarboxylation of histidine by the enzyme histidine decarboxylase (which requires the coenzyme pyridoxal 5′-phosphate)**

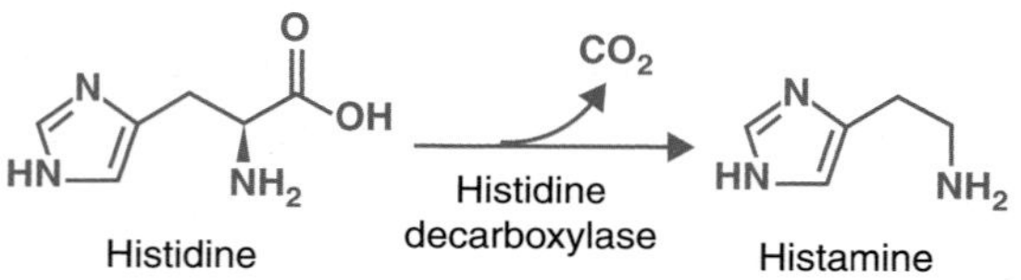

Histidine   Histidine decarboxylase   Histamine

BOX 7.1

Are you allergic to shellfish?

Relatively high levels of histamine can be found in a variety of foods, such as fish, including shellfish, chicken, tomatoes, spinach, and chocolate. Food and drinks in which bacterial, yeast, or mould growth has occurred (wine, beer, cheese, some processed meats, relishes, and pickles, for example) may also contain high levels of histamine. Normally, this histamine is broken down by the enzyme diamine oxidase produced in the small intestine, and so does not affect the consumer. Some individuals with abnormally low levels of this protective enzyme, however, may experience allergic-like symptoms when they eat an excessive quantity of histamine-containing food. Without adequate levels of diamine oxidase, histamine may be absorbed from food and reach plasma levels that are sufficient to cause allergic-like symptoms.

> Section 4.2 'Basic concepts of pharmacodynamics', within Chapter 4 of this book introduces the various different types of receptor, including G-protein-coupled receptors.

There are different types of histamine receptors present in specific tissues. They include:

- H_1 receptors, which are found in the smooth muscle cells, and when excited by histamine cause the muscles to contract. They are also found in the smooth muscle cells associated with blood vessel walls, particularly in large veins. Stimulation of these receptors by histamine induces vasodilatation.

- H_2 receptors, which are found in the gastric glands of the stomach and, when activated, stimulate the production of gastric juice. H_2 receptors are also found on cardiac muscle, which may be stimulated by histamine, and on the surface of mast cells, where the release of histamine may be regulated by a negative feedback system.

- H_3 receptors, which are found in the brain. These are found on presynaptic nerve endings and when activated are able to inhibit the release of neurotransmitters such as acetylcholine, noradrenaline, and dopamine.

- H_4 receptors were discovered relatively recently and are thought to participate in the development of some of the components of inflammatory reactions. Their stimulation seems to give similar outcomes to those seen with H_1 receptors. They may also promote the release of neutrophils from their source in the bone marrow.

> **KEY POINT**
>
> Histamine is an autocoid that is involved in both physiological and pathological processes. There are several types of histamine receptor, with H_1 receptors on blood vessels and H_2 receptors in the stomach being of particular importance.

> **SELF CHECK 7.1**
>
> What are the effects of histamine on peripheral blood vessels?

Effects of histamine on the body

As histamine is found in many tissues around the body it has an effect on a number of different body systems. Some of these are considered in this section.

Gastrointestinal tract

Histamine induces contraction of intestinal smooth muscle and stimulates gastric juice secretion. As well as having a direct effect on *parietal* (hydrochloric acid-secreting) and *oxyntic* (pepsin-secreting) cells, histamine also enhances the secretory activity of other stimulants such as the neurotransmitter *acetylcholine* and the hormone *gastrin*.

This effect of histamine is very profound and in the absence of histamine the secretory activity of the stomach is substantially reduced. The reduction of secretion into the stomach is a very useful goal in a variety of clinical situations, including dyspepsia and

peptic ulceration. It is also a target in cases where ulcers may have been provoked by the use of non-steroidal anti-inflammatory drugs (NSAIDs).

Airways

Air breathed in through the nose or mouth passes through a series of ever-diminishing tubular airways to get to the alveoli—structures where gases are exchanged between air and the blood. These tubular airways include: the trachea (which divides into a left and a right bronchus), the small bronchi (which branch from the bronchus) and the even smaller bronchioles (branching off the bronchi). These are all surrounded by a layer of smooth muscle which can be stimulated to contract by histamine. Contraction of this bronchial smooth muscle reduces the size of the airways causing a reduction in the volume of air that can get to the alveoli.

The spontaneous release of large amounts of histamine within the body can have very profound effects on the ability of the lungs to function adequately. Some individuals, notably those who suffer from asthma, are very sensitive to this constricting effect of histamine.

Uterus

The walls of the uterus include a significant amount of smooth muscle. This muscle may also be stimulated to contract by histamine. This is a powerful effect in some species but the human uterus is less sensitive to the action of histamine. Nevertheless, in extreme circumstances severe consequences may be induced by histamine, as described in Box 7.2.

Nerves

Histamine is a powerful stimulant of nerve endings, with sensory nerves being particularly susceptible. The injection of small quantities of histamine in the skin may give rise to the sensation of itching (*pruritus*), and larger quantities may induce a pain response, though this is generally only mild in severity. These effects may be inhibited by selective antagonism of H_1 receptors with traditional antihistamine drugs such as mepyramine or chlorphenamine, which are used in the alleviation of the symptoms associated with insect bites and urticaria.

The role of histamine in pregnancy

Whether histamine has a role in pregnancy remains an open question. It is clear that the placenta (the vascular exchange structure that links the maternal and fetal circulations) produces high levels of diamine oxidase during pregnancy, and it is assumed that this protects the developing fetus from any excess histamine present in the maternal circulation. In addition, the placenta also exhibits a high level of histidine decarboxylase, which tends to suggest that histamine may have a significant role during pregnancy. This may be as a vasodilator, ensuring a good supply of blood to the placenta. It is also possible that the growth-promoting properties of histamine may be important to embryological development.

Cardiovascular system

The effects of histamine on the circulatory system are complex and, indeed, complicated by reflexes that attempt to reverse some of the histamine-induced changes. If high levels of histamine are injected in to the body or if large amounts of stored histamine are released, there is usually a very significant drop in blood pressure as histamine causes many peripheral blood vessels to relax. The heart rate may consequently rise owing to reflex activity in an attempt to raise the blood pressure. There may also be a marginal rise in heart rate because of a direct effect on H_2 receptors present in the heart. The fall in blood pressure caused by histamine activity can be very serious and, in some circumstances, life-threatening.

Histamine possesses another property that affects blood vessels—it increases the permeability to plasma protein of small blood vessels, notably venules—and may thus exacerbate the fall in blood pressure by inducing a decrease in blood volume as well. See Box 7.3 for more information on blood vessel permeability.

KEY POINT

Histamine released from mast cells can cause a dramatic fall in blood pressure, and constriction of airways that makes breathing very difficult.

Inflammation

Histamine may be involved in a variety of acute inflammatory reactions, particularly in the early development stage of the reaction. The most obvious examples of inflammatory reactions in which histamine is involved are the *type I allergic reactions*, such as those seen in hay fever (acute allergic rhinitis). In this reaction the upper respiratory mucosa of a sensitive individual becomes inflamed as the subject's mast cells are induced to release histamine following interaction between pollen (the **antigen**) and specific **antibodies** (immunoglobulin E, IgE) present on the surface of the mast cell. An increased supply of blood is drawn to the affected mucosa and, following a migration of protein from blood through the expanded gaps between endothelial cells, as described in Box 7.3, water moves from the blood into the nasal passages giving rise to what we may observe as a runny nose.

Therapeutic agents suppressing the action of histamine

There are two clinically relevant types of histamine receptor antagonists, H_1 and H_2.

H_1-receptor antagonists (antihistamines)

The antihistamines include some of the oldest completely synthetic drugs. The earliest examples have been available since the 1940s. They are all competitive antagonists of histamine at H_1 receptors. They also commonly exhibit some anticholinergic activity at **muscarinic receptors**. Their most obvious use is for the treatment of hay fever for which they are undoubtedly highly effective. It is important to note, however, that they are often used inappropriately and it is necessary to have an understanding of their **pharmacokinetic** properties to be sure that they are used in the most effective way.

Examples of the older generation antihistamines are: alimemazine, chlorphenamine, clemastine, cyclizine, cyproheptadine, and promethazine. Traditional antihistamines are generally quite sedative, with promethazine being particularly likely to induce drowsiness. Of course, some sedation may be quite helpful in dealing with troublesome symptoms like itchy skin at

BOX 7.3

Fluid equilibrium

Ordinarily, low molecular weight substances such as water molecules, and small solutes such as sodium and chloride ions, as well as glucose molecules, are able to move freely between the blood and the tissue fluid in which the cells and tissues are bathed. This movement occurs through tiny gaps between adjacent cells lining the blood vessels, known as *endothelial cells*. As the pressure of the blood inside blood vessels is much higher than the pressure of the fluids outside the blood, there is an inevitable tendency for these small molecular weight substances to move from the blood into the tissues.

The water that moves out of the blood under the influence of positive blood pressure is ultimately drawn back into the blood owing to the osmotic potential of the protein present in solution within the blood. This effect induced by the protein in blood is often referred to as *oncotic* pressure. In health, this osmotic pressure exactly counterbalances the **haemodynamic pressure** of the heart pumping and a dynamic fluid equilibrium is established and maintained, as shown in Figure 7.2(A).

If the permeability of small blood vessels is changed such that some plasma protein is able to leak out into the tissues, the equilibrium point will change and some fluid will be retained outside the vessels in the tissues, as indicated in Figure 7.2(B). This extravascular protein will exert an osmotic pressure, thus retaining a proportion of the fluid in the tissues. If this fluid retention persists then one or two very significant changes may be observed. In a situation where the tissue boundary remains intact and is relatively impermeable (as would be the case, for example, in a sprained ankle) there will be an accumulation of fluid which may be observed

as a swollen joint. This accumulated fluid is known as *oedema*. If the changes occur where the skin is damaged or in tissues with more flimsy barriers, such as the nasal mucosa, then fluid will be seen to leak from the site. In conditions like the common cold or hay fever the nasal exudate observed is the leaking of this oedematous fluid.

FIGURE 7.2 Dynamic fluid equilibrium is established in normal tissues (A), and no extravascular protein is leaked. In inflamed tissues (B) the accumulation of protein in extravascular spaces facilitates the movement of water into the tissues, leading to oedema or the secretion of exudates

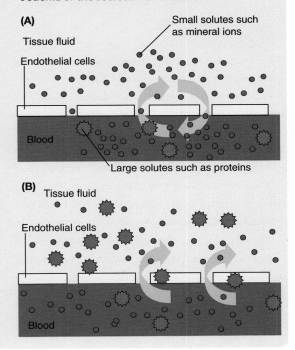

night. Of the older short-acting antihistamines, chlorphenamine, by reputation, is the least troublesome with respect to drowsiness.

Some of the newer antihistamines, introduced after 1970, offer the advantage of longer duration of action and less sedation. The reduced sedation is attributed to their molecular structure making them less likely to cross the **blood–brain barrier**. Examples of the newer antihistamines include: cetirizine, a fairly

short-acting drug molecule but quick to exert effect, and loratadine, a longer-acting drug molecule which takes several days to reach therapeutic levels. Case Study 7.1 looks into the use of these antihistamines. Others include: desloratadine, fexofenadine, levocetirizine, mizolastine, and rupatadine. Figure 7.3 shows the structures of some of these antihistamines.

Histamine (and acetylcholine) can stimulate the vestibular nuclei in the medulla oblongata (base of

FIGURE 7.3   The structure of the antihistamines desloratadine, alimemazine, and cetirizine

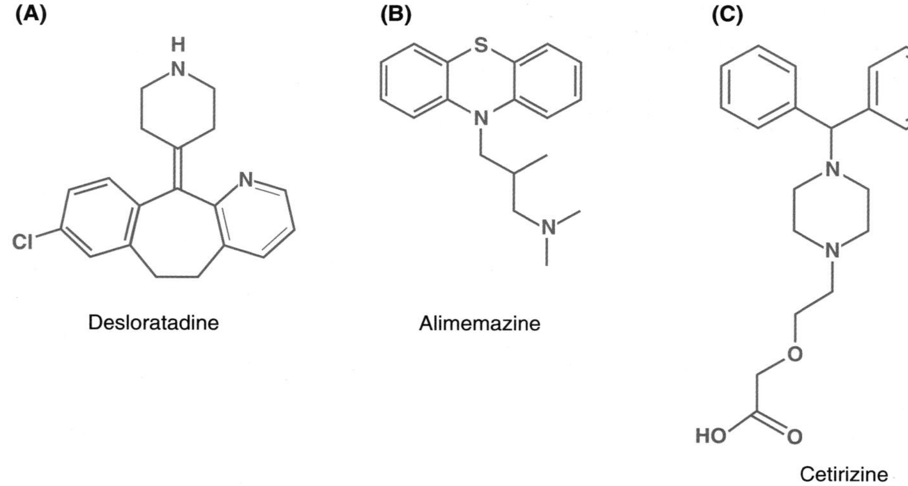

(A) Desloratadine

(B) Alimemazine

(C) Cetirizine

CASE STUDY 7.1

Sarah was complaining to her local pharmacist of a runny nose and itchy eyes. She is a hay fever sufferer and experiences symptoms of variable severity at this time of year. She explained that she always takes a loratidine tablet (10mg) when symptoms occur but that she did not always find this to be effective. The pharmacist provided an alternative preparation of cetirizine (10mg), with the advice that she should take one immediately and repeat the dose four-hourly, whilst symptoms persisted. He further advised her to start the loratidine tablets again when symptoms were less troublesome, but to ensure that she took a tablet every day throughout the hay fever period, even when she was symptom free.

REFLECTION QUESTIONS

1. Why was the loratadine tablet not effective in managing Sarah's hay fever symptoms?

2. Why did the pharmacist give Sarah certirizine tablets to treat the symptoms?

3. Why do hay fever suffers not all suffer their symptoms at the same time?

Answers

1 This is because of the pharmacokinetic profile of the loratadine. A single dose does not achieve an effective therapeutic level of the drug. It takes several days. In consequence it is not a useful drug for responding to symptoms. On the other hand, if it is taken just once daily, after a few days an effective level is reached and it is possible to sustain this level of the drug in the body indefinitely. This is an excellent approach for hay fever sufferers as it is perfectly feasible to prevent (or at least reduce) the symptoms of hay fever over the sustained period of the pollen season. The key to this approach is undoubtedly to keep taking the tablets, even in the absence of symptoms. Of course, without the prompt of symptoms, this can be difficult to remember.

2 Short-acting antihistamines like cetirizine (and chlorphenira-mine) can provide effective relief very quickly. These drugs are very appropriate for a rapid response to symptoms. A tablet is taken at frequent intervals throughout the day, whilst symptoms persist. An effective therapeutic concentration will be reached within an hour. Taking fast but short-acting medicines is a really good way to respond to symptoms but is troublesome as the dose needs to be repeated many times a day and this is why Sarah was advised to start taking the loratadine again every day, even when she is symptom free.

3 Because individuals are not all affected by pollen from the same species. Pollen from deciduous trees tends to be released into the air between March and May, whilst grass pollen tends to be released in June and July.

the brain), which can lead to activation of the vomiting centre. Antihistamines are thus effective in the treatment of nausea and vomiting, particularly motion sickness. Additionally, in patients with Parkinsonism, antihistamines may be used to decrease stiffness and reduce tremor.

H_2-receptor antagonists

H_2-receptor antagonists became generally available in the mid 1970s and are very useful in the treatment of dyspepsia, reflux oesophagitis (heartburn), and peptic ulceration. Cimetidine is a competitive inhibitor of H_2 receptors and is very effective; in typical clinical trials, 95% of subjects with peptic ulcers would be symptom-free within a few weeks of treatment. The drug is very selective and has few side effects, such as temporary male infertility and inhibition of the metabolism of some drugs. In practice, although cimetidine only affects drugs metabolized via the cytochrome P450 route, it is safer to avoid this drug where other drugs are also being taken. This problem is not shared by the newer H_2-receptor antagonists: ranitidine, nizatidine, and famotidine. See Box 7.4 for more information about the discovery of cimetidine.

While the H_2-receptor antagonists are very effective in dealing with the symptoms of peptic ulceration, they are less effective in curing ulcers. This is not entirely surprising as they are only used to reduce gastric secretion and do not target the cause of the ulcer (which may have a microbiological origin in many cases). Nevertheless, a rise in the pH of the stomach contents is likely to facilitate new epithelial growth in the base of the ulcer, allowing a degree of recovery.

Within a decade of the introduction of H_2-receptor antagonists for treating peptic ulcers, three other new types of similarly effective drugs with different mechanisms were introduced to clinical practice: the

BOX 7.4

The discovery of cimetidine

The development of the first clinically successful H_2-receptor antagonist by the Nobel Laureate James Black and his team marked a turning point in the approach to drug design. The team set about modifying the histamine molecule in an attempt to find a drug that would inhibit the stimulant effect of histamine on acid secretion in the stomach. This rational approach to drug design (which James Black similarly applied in the creation of propranolol) represented a huge step forward in the field of drug development. This work culminated in the launch of the H_2-receptor antagonist cimetidine. Prior to the launch of cimetidine, the drug of first choice for peptic ulceration was a liquorice extract, which was not very effective and surprisingly toxic.

gastroselective **antimuscarinic drug** pirenzepine (now withdrawn), **proton-pump inhibitors** such as omeprazole, and the **prostaglandin** analogue misoprostol. All are very safe and similarly effective in the treatment of peptic ulcers.

KEY POINT

Therapeutic agents targeting histamine receptors are antagonists and their clinical application depends on their selectivity for the receptor types. They are commonly used to treat hay fever and pruritus (H_1-receptor antagonists) and dyspepsia, heartburn, and peptic ulceration (H_2-receptor antagonists).

SELF CHECK 7.2

What is hay fever and how are antihistamines used to treat it?

7.3 Serotonin

Serotonin is another example of a locally acting hormone, although interest in serotonin is more focused on its neurotransmission role. The most obvious roles of serotonin, outside the brain, are its involvement in peristalsis and platelet aggregation. Serotonin is also known *as 5-hydroxytryptamine*, which is its chemical name, often shortened to *5-HT*, as well as *enteramine* and *vasotonin*. It is a biologically active amine, as is histamine, and is found in both the plant and animal kingdoms. It is a frequent component of venoms but in humans it is found in platelets and the intestinal mucosa. It is also present at a variety of sites in the brain.

Serotonin is synthesized by the hydroxylation of the indole ring of the amino acid tryptophan by the enzyme tryptophan 5-hydroxylase, followed by decarboxylation by aromatic-L-amino-acid decarboxylase, as shown in Figure 7.4. It is stored in *enterochromaffin cells* (so named by histologists who discovered them in intestinal tissue, the *enteron*, and observed that they had an *affinity* for *chromium*-containing stains). Platelets do not synthesize serotonin but actively absorb it to quite high concentrations.

FIGURE 7.4 Serotonin synthesis occurs by the hydrolysis and subsequent decarboxylation of the amino acid tryptophan

Tryptophan

Tryptophan
5-hydroxylase

5-Hydroxy-tryptophan

Aromatic-L-amino-
acid decarboxylase

Serotonin

5-Hydroxy-tryptamine
(5-HT)

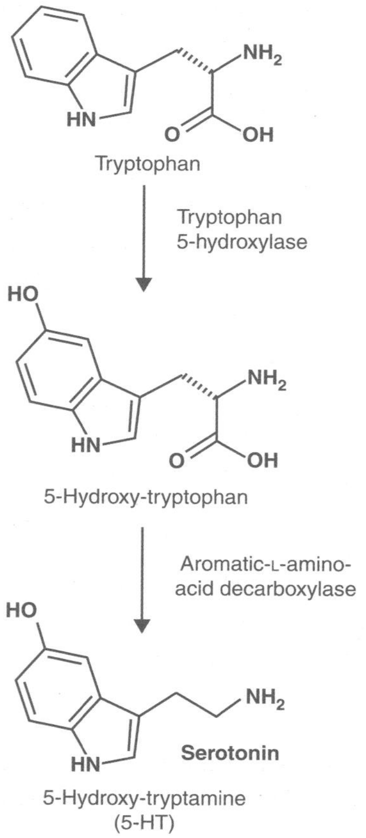

Following release, serotonin is metabolized via oxidation by monoamine oxidase to yield 5-hydroxy-indole acetaldehyde and then further oxidized by aldehyde dehydrogenase to 5-hydroxyindoleacetic acid (5-HIAA). Daily serotonin turnover can be monitored by an analysis of the 24-hour urinary excretion of 5-HIAA.

There are more than ten different serotonin receptors. The main types of clinical interest are:

- 5-HT_1, which are inhibitory receptors on nerve endings.
- 5-HT_2, which are stimulatory receptors on smooth muscle.
- 5-HT_3, which are stimulatory receptors on nerve endings.
- 5-HT_4, which are stimulatory receptors on motor nerves in the gastrointestinal system.

Each of these categories of receptor may be further subdivided. A suffix letter is added after the number to denote this. While beyond the scope of this chapter, it is appropriate to mention that there are many different subtypes of serotonin receptor present in the brain and there is great interest in the manipulation of these receptors to treat psychiatric and neurological problems. Areas of interest include sleep, emotion, sexual activity, and appetite suppression.

KEY POINT

There are at least ten different types of serotonin receptor found in the body, the clinically most important of which are 5-HT_1, 5-HT_2, 5-HT_3, and 5-HT_4.

Effects of serotonin on the body

Serotonin affects a number of different organs and body systems, some of which are discussed in this section.

Gastrointestinal tract

More than 90% of the body's serotonin is found in the gastrointestinal system, principally in enterochromaffin cells. When released from these cells, serotonin stimulates 5-HT_1 and 5-HT_3 receptors on local nerves, increasing the smooth muscle tone of the gastrointestinal tract and facilitating peristalsis.

Tumours of **argentaffin cells** (carcinoid syndrome) can secrete larger than normal quantities of serotonin. This results in unpleasant symptoms such as diarrhoea and bronchoconstriction.

Unlike histamine, serotonin has little effect on gastrointestinal secretory activity. It is, however, effective at stimulating the vomiting reflex by both peripheral and central effects.

Airways

Serotonin is a powerful bronchoconstrictor, though it seems unlikely that this activity is very important in normal function of the respiratory system in humans.

Cardiovascular system

Serotonin has a number of effects on the cardiovascular system. It is a very powerful constrictor of large blood vessels, an effect brought about by stimulation of 5-HT$_2$ receptors on the vascular smooth muscle. Serotonin dilates coronary and skeletal muscle blood vessels. This pattern of action is very similar to that caused by adrenaline (also known as epinephrine).

Like adrenaline, serotonin has positive **inotropic** and **chronotropic** effects on the heart, but these effects are much weaker than are seen in response to adrenaline. Reflex **bradycardia** can also be seen with serotonin which is induced by the stimulation of **chemoreceptors**.

Serotonin may also cause vasodilatation by indirect mechanisms. It can inhibit the release of noradrenaline from sympathetic nerves and can stimulate the release of the vasodilator nitric oxide from endothelial lining cells.

Dilation of blood vessels in the brain may give rise to headaches and may also be a contributory factor in migraine. These vessels can be constricted by the action of serotonin on 5-HT$_1$ receptors. This property lends itself to the treatment of such disorders.

Nervous system

Serotonin can stimulate sensory nerve endings, causing the sensation of itching. More powerful stimulation may induce pain. This effect is mediated by 5-HT$_3$ receptors.

Platelets

Serotonin can promote aggregation of platelets when they are called to plug damaged blood vessels. This response is mediated by 5-HT$_2$ receptors. Once aggregation starts, the platelets release more serotonin from their stores. This locally released serotonin can modify the diameter of the blood vessel in different ways, depending on the prevailing circumstances. Direct stimulation of 5-HT$_2$ receptors on vascular smooth muscle will cause constriction, which will help to reduce blood loss from a damaged vessel.

> **SELF CHECK 7.3**
>
> What is the role of serotonin in headaches?

Drugs acting on serotonin receptors

Several drugs which interact with serotonin systems are available clinically. They can be divided into several categories and are discussed in this section.

Serotonin antagonists

Many drugs blocking other substances also antagonize serotonin receptors. Antihistamines and α-adrenoceptor antagonists generally exhibit some antiserotonin activity as well. Examples of serotonin antagonists include:

- Cyproheptadine, which is an antihistamine with some 5-HT$_2$ antagonist activity. It also has significant antimuscarinic activity. It is used to treat allergic conditions such as hay fever and urticaria.

- Ketanserin, which antagonizes 5-HT$_{1c}$ and 5-HT$_2$ receptors. It is highly selective and blocks platelet aggregation. It is mainly used in the management of hypertension.

- Methysergide, which is an antagonist of 5-HT$_2$ receptors and may be used to treat vascular headaches. Its use is limited, however, by its side effects.

Serotonin agonists

Some drugs are highly selective stimulants of serotonin receptor subtypes and are used in a very wide range of neurological and psychiatric clinical situations. Examples include:

- Buspirone, which stimulates 5-HT$_{1a}$ receptors. It has anxiolytic activity and is effective in alleviating anxiety states.

- Dexfenfluramine, which stimulates 5-HT receptors and suppresses appetite. It has been withdrawn from clinical use, however, owing to the risks of serious adverse effects.

- Sumatriptan (a 'triptan'), which stimulates 5-HT$_{1b/1d}$ receptors and is an effective treatment of migraine. Similar drugs used to treat migraine include: almotriptan, eletriptan, frovatriptan, naratriptan, and zolmitriptan.

- Metoclopramide, which stimulates dopamine receptors as well as 5-HT$_4$ receptors. It also antagonizes 5HT$_3$ receptors and has proved to be a very useful drug, both as an antiemetic and a stimulant of gastric emptying.

Selective serotonin reuptake inhibitors

Serotonin activity in the brain is closely related to mood, and lower turnover rates of serotonin can be associated with depression. Selective serotonin reuptake inhibitors (SSRIs) act by enhancing concentrations of serotonin available within the brain. This is achieved by the simple expedient of inhibiting the natural serotonin removal process. SSRIs are in clinical use for the treatment of depression. Examples include citalopram, escitalopram, fluoxetine, fluvoxamine, paroxetine, and sertraline. It is inevitable that a general inhibition of reuptake will enhance the activity of serotonin throughout the body. Side effects on the gastrointestinal system are common and include diarrhoea.

Ergot alkaloids

The ergot alkaloids include a number of substances that stimulate serotonin receptors. Their role in the history of medicine, and in pharmacology in particular, is interesting. Ergot is a fungus which can contaminate wheat and other cereals. Throughout the Middle Ages, outbreaks of what came to be called St Anthony's fire were common. Profound, painful peripheral vasoconstriction (often leading to gangrene) together with a range of psychoses were the most obvious symptoms, with abortion the frequent consequence in pregnancy. The alkaloids are generally derivatives of lysergic acid and many of them are effective stimulants of 5-HT receptors. They also stimulate α-adrenergic receptors and dopamine receptors. Extracts include ergometrine, used to treat excessive postpartum bleeding; ergotamine, used to treat migraine; and methysergide (mentioned under 'Serotonin antagonists' in this section).

7.4 Kinins

Originally discovered in the first half of the 20th century and characterized in the 1960s by Elliott, Lewis, and Horton, kinins are another interesting example of locally acting hormones. They are very different substances to the amines already considered (histamine and serotonin) as they are short-chain peptides. *Bradykinin*, the most commonly encountered kinin, is a peptide containing nine amino acids (a nonapeptide). Its amino acid sequence is: arginine, proline, proline, glycine, phenylalanine, serine, proline, phenylalanine, arginine. *Kallidin,* also found in the human body, is a decapeptide (ten amino acids). At equimolar concentrations, kinins are much more potent than their amine counterparts. Unlike histamine and serotonin they are not stored by the body but synthesized from larger molecules as and when they are required. They are, however, very rapidly inactivated after release.

Kinins may be generated by a variety of different mechanisms. In blood, they are produced by the action of an enzyme called *kallikrein* on a plasma globulin protein substrate known as a *kininogen*. Kallikrein itself is synthesized as an inactive precursor, kallikreinogen, though this is more commonly referred to as prekallikrein. It is activated in plasma by a number of different stimuli, for example dilution of blood (as may occur during haemodialysis). Contact between prekallikrein and exposed collagen fibres (which may happen following injury-induced damage to blood vessels) may also activate the *kallikrein–kinin system*.

The kallikrein–kinin system is involved with blood clotting processes. Activation of the fibrin-generating process is likely to be associated with kinin generation. Hageman factor (factor XII in the blood clotting cascade) activation is intimately associated with the activation of kallikrein. Hageman factor activation will promote the activation of both the clotting cascade and kallikrein.

 You can read more about the process of blood coagulation in Section 9.6 'Haemostasis', within Chapter 9 of this book.

Bradykinin activity is terminated by hydrolysis. In humans, bradykinin is broken down by a variety of different enzymes, generally referred to as *kininases* (not to be confused with kinases, which are enzymes involved in the transfer of phosphate groups). One of these, angiotensin-converting enzyme (ACE), can split the bradykinin peptide between amino acids 7 and 8. An aminopeptidase can remove the arginine amino acid at the *N*-terminal end, while a carboxypeptidase can remove the arginine amino acid at the *C*-terminal end of the molecule. Any of these changes will terminate the biological activities of bradykinin.

Bradykinin stimulates G-protein-coupled receptors, two subgroups of which have been identified: B_1 and B_2. Stimulation of the B_1 receptor gives rise to a range of responses generally promoting inflammatory activities. The stimulation of the B_2 receptor is thought to provide some cytoprotection to cardiac muscle following ischaemic damage, by promoting local vasodilatation. Expression of the B_1 receptor is induced by tissue damage, and pharmacological agonists and antagonists of this receptor seem to have mostly noxious effects. The constitutively expressed B_2 receptor, however, when activated exerts mostly beneficial actions.

> ## KEY POINT
>
> Kinins are potent biologically active local hormones. They are not stored but generated as required.

Effects of kinins on the body

Kinins have been shown to induce a number of significant changes in tissues. They are powerful inducers of smooth muscle contraction and, in asthmatics, may promote particularly troublesome bronchoconstriction. Kinins are also powerful vasodilators and, as is seen with histamine, promote an increase in the permeability of small blood vessels.

When the temperature of the body rises it responds by relocating blood to the skin, where heat loss may occur by radiation. The relaxation of the blood vessels in the skin is not induced by cholinergic nerves of the parasympathetic nervous system but may be encouraged by the local generation of kinins. Heat loss is further enhanced by the secretion of sweat onto the surface of the skin. The evaporation of sweat is an endothermic process and evaporation of sweat significantly enhances the reduction of the body's temperature. Bradykinin levels in and around sweat glands rise during sweating and it is likely that the functional vasodilatation required for sweat gland activity is also induced by the local generation of this kinin.

 Temperature regulation by the body is further addressed in Section 8.3 'Major homeostatic mechanisms in the human body', within Chapter 8 of this book.

Kinins have been regarded as mediators of the inflammatory response. Their properties (vasodilatation, vascular permeability changes, and pain promotion) are certainly consistent with this concept. They are also believed to have a significant role in the pathology of acute pancreatitis.

Kinins may also promote coughing. ACE inhibitors such as captopril, which can be used to treat hypertension, may also inhibit the breakdown of kinins. The consequent rise in their tissue levels may account for the development of a dry cough in some patients taking ACE inhibitors.

Kallikrein activity can be inhibited by the protease inhibitor *aprotinin*. There has been significant interest in this potential therapeutic agent for the treatment of inflammation and acute pancreatitis. Its effectiveness in treatment has not been adequately substantiated, however, and it is no longer listed in the British National Formulary (BNF).

A range of experimental kinin antagonists have been tested in a number of situations. These are biologically inactive nonapeptides, similar to kinins, that have a substituted amino acid (usually a D-aromatic amino acid) in place of proline in position 7. Clinical trials with inactive kinin analogues have demon-

strated some effects in inflammatory disease, asthma, and rhinitis. The benefits, however, have been somewhat disappointing.

7.5 Eicosanoids

Eicosanoids are bioactive molecules derived by enzyme action following the release of fatty acids from their membrane-bound source. Eicosanoids are not generally stored and, like kinins, are synthesized on demand. There are two major enzyme systems that lead to the production of active products:

- The *cyclooxygenase* (COX) enzymes produce *prostaglandins* (PG) or *thromboxanes* (TXA).

- The *lipoxygenase* enzymes generate *leukotrienes* (LT).

The rate-limiting factor for eicosanoid production is substrate availability. The principal substrates for eicosanoid synthesis are the long-chain polyunsaturated fatty acids *arachidonic acid* and *γ-linolenic acid*, which are shown in Figure 7.5.

Arachidonic acid is the more abundant of the two substrates and is released from cell membrane phospholipid stores by phospholipase enzymes such as phospholipase A2. Arachidonic acid is a 20-carbon straight-chain fatty acid that contains four double bonds. Arachidonic acid metabolism leads to production of the '2' series eicosanoids, for example prostaglandin E_2 (PGE$_2$) and thromboxane A_2 (TXA$_2$). γ-Linolenic acid is a carbon straight-chain fatty acid that contains three double bonds. γ-Linolenic acid

metabolism leads to production of the '1' series eicosanoids such as PGE$_1$.

All cells are able to produce an eicosanoid. When stimulated, cells release fatty acids from membrane stores and, once available, the fatty acids are metabolized by whichever enzyme pathway is available in the cell to generate the characteristic product for the cell type.

KEY POINT

The eicosanoid product made by a cell depends partly on the substrate available and partly on the enzyme system present in the particular cell.

Cyclooxygenase

Cyclooxygenase is the enzyme which generates prostaglandins from available unsaturated fatty acids. PGE$_2$ is an eicosanoid produced by cyclooxygenases, and its structure is shown in Figure 7.6. There are two identifiable cyclooxygenase enzyme systems exhibiting significantly different properties:

- Cyclooxygenase 1 (COX1) is expressed in most tissues and has a constant presence. It is deemed to be

FIGURE 7.5 The structure of arachidonic acid and γ-linolenic acid

Arachidonic acid

γ-Linolenic acid

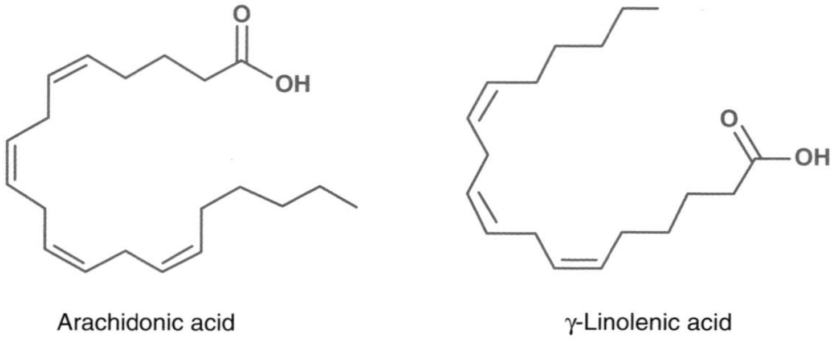

FIGURE 7.6 Structure of prostaglandin E$_2$ (PGE$_2$)

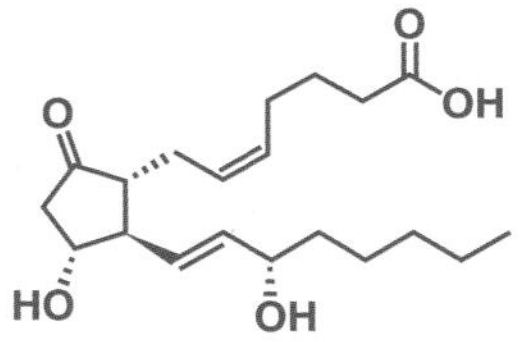

responsible for producing prostaglandins that are involved in cellular homeostasis. It may respond to the prevailing physiological activity of groups of cells, enabling a very fine level of control over the processes in which those cells are engaged. Suggested roles include the maintenance of the microvascular integrity of the gastrointestinal tract, regulation of cell division, regulation of mucus production, and the maintenance of renal vascular tone.

- Cyclooxygenase 2 (COX2) is not generally available and is only expressed at sites of inflammation. This inducible enzyme is thought to be responsible for the pain associated with inflammation.

Lipoxygenases

The lipoxygenases are found principally in lung tissue and white blood cells, and their derivatives such as mast cells and platelets. A key enzyme is arachidonate 5-lipoxygenase, more commonly known simply as 5-lipoxygenase. When cells are activated, this enzyme becomes linked to a membrane-bound accessory protein, the *5-lipoxygenase-activating protein*. This enzyme converts available substrates, such as arachidonic acid, to leukotriene B4 (LTB$_4$). Leukotriene C4 (LTC$_4$) is produced by the addition of the tripeptide glutathione to LTB$_4$ by the enzyme glutathione *S*-transferase. LTC$_4$ may subsequently release these amino acids sequentially to progressively give rise to LTD$_4$, which contains a dipeptide, and LTE$_4$, which contains a single amino acid. The loss of this final amino acid results in LTB$_4$.

Figure 7.7 shows the structure of the different leukotrienes. Leukotrienes are generally present as a mixture of these various molecules. They were originally demonstrated to be products of anaphylaxis and given the name *slow-reacting substance of anaphylaxis* or SRSA. It was not until half a century later that their chemical composition was identified.

Effects of eicosanoids on the body

The different eicosanoids have varying effects on the body.

Prostaglandins in the stomach

Prostaglandin receptors are present on the parietal cells. Stimulating these receptors reduces the activity of the enzyme adenylate cyclase. This results in

141

FIGURE 7.7 The structure of leukotrienes B$_4$, C$_4$, E$_4$, and D$_4$

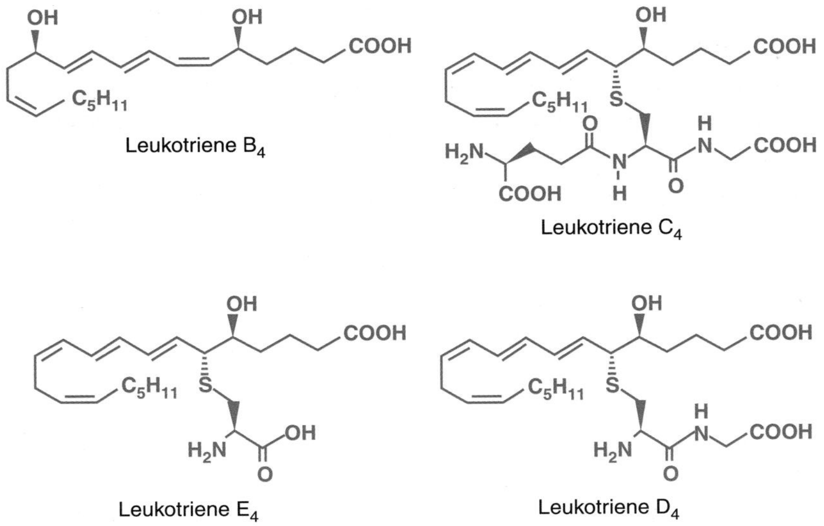

Leukotriene B$_4$

Leukotriene C$_4$

Leukotriene E$_4$

Leukotriene D$_4$

a lowering of the level of the intracellular second messenger 3′,5′-cyclic adenosine monophosphate (cAMP), leading to a reduction of hydrochloric acid output from the parietal cells. Prostaglandins may, therefore, have a role in moderating the pH of the stomach. The vasodilatory properties of prostaglandins may promote blood flow through actively secreting gastric tissues. In addition, prostaglandins stimulate the secretion of the protective mucus that coats the gastric mucosa (innermost layer of the stomach) and promote the proliferation of the epithelial lining cells.

Eicosanoids and blood clotting

PGI_2, also known as prostacyclin, is produced by the lining cells of blood vessels (endothelium). It acts to inhibit platelet adhesion and promotes vasodilatation. It may thus prevent excess blood clot formation. Virtually opposite effects are seen with thromboxanes, produced by platelets themselves; TXA_2 promotes platelet aggregation and vasoconstriction. The very different properties of the two eicosanoids provide a good insight into the concept that these substances may exert control over a system by variation in their relative concentrations. Thromboxanes promote vasoconstriction and platelet aggregation, which are very important activities for dealing with damaged blood vessels, helping to limit the loss of blood. Excessive platelet aggregation, however, may lead to the development of unwanted thrombosis. The counter-activities of PGI_2 may help to limit the occurrences of such unwanted critical events.

Eicosanoids and the airways

Both PGI_2 and PGE_2 induce relaxation of respiratory smooth muscle, causing an increase in the diameter of the airways and allowing an increase in air passage. $PGF_{2\alpha}$ and TXA_2 have been shown to cause the exact opposite effects and both can cause powerful contraction of respiratory smooth muscle. Their actions will lead to a restriction in the amount of air that may pass in and out of the lungs. Leukotrienes are also profound bronchoconstrictors. The balance between bronchoconstrictor and bronchodilator eicosanoids may be a significant factor in the fine control of the airways. Interference with this balance has the potential for untoward effects.

It is evident that many asthmatics are very sensitive to the bronchoconstrictor activities of eicosanoids. About 30% of people with asthma react severely to the ingestion of NSAIDs such as aspirin. Aspirin is an inhibitor of the cyclooxygenase system and by blocking this pathway it effectively increases the availability of arachidonic acid for metabolism by the lipoxygenase pathway. The inevitable increase in bronchoconstricting leukotrienes probably explains the reaction seen in aspirin-sensitive asthmatics.

Leukotrienes and inflammation

Leukotrienes are potent chemotactic agents for white blood cells. They also promote cell adhesion to the lining of blood vessel walls, which is an essential prerequisite for the migration of white blood cells from the blood into the surrounding tissues. They also promote degranulation of mast cells and promote an increase in the permeability of small blood vessels to protein, leading to the development of oedema. Leukotrienes may therefore be significant mediators in the development of inflammatory reactions.

Prostaglandins and blood vessels

Prostaglandins are very potent vasodilators. Their effect is very long lasting, certainly exceeding the duration of their presence. Furthermore, their dilating effect is enhanced by the presence of other substances such as histamine and kinins. Prostaglandins also promote the increase in blood vessel permeability induced by histamine and kinins. This is an enhancing effect rather than a direct effect.

PGD_2, produced peripherally by mast cells, is the primary mediator of the vasodilatation that occurs following the ingestion of nicotinic acid (a vitamin B derivative, also known as niacin, that may be used to lower cholesterol levels in the blood). The PGD_2 antagonist laropiprant is used in combined preparations with nicotinic acid to limit the vasodilator side effect (manifested as flushing) associated with the use of the vitamin.

Prostaglandins and the kidney

Prostaglandins levels are increased in the kidney when blood pressure falls. This induces vasodilatation within the renal blood vessels, maintaining blood flow.

Prostaglandins potentiate the pain-causing effects of other peripherally acting inducers of pain. Unless injected into the tissues at very high concentrations (higher than found in normal physiological circumstances) prostaglandins do not cause pain directly.

PGE_1 and PGE_2 are found at higher than normal concentrations in the central nervous system during fever (increased body temperature). Interleukin 1 (IL-1), produced by macrophages in response to the presence of microorganisms, promotes the synthesis of PGE_2. IL-1, tumour necrosis factor-α (TNF-α) and interleukin 6 (IL-6) are important endogenous pyrogens involved in the process of adjusting the thermoregulatory set-point in the hypothalamus (the body's thermostat) to induce pyrexia (fever response).

SELF CHECK 7.4

How do the effects of TXA_2 differ from those of PGI_2?

Effect of NSAIDs on prostaglandins

NSAIDs have been shown to inhibit synthesis of prostaglandins in most tissues. Common NSAIDs include aspirin, ibuprofen, diclofenac, ketoprofen, naproxen, and piroxicam. Cyclooxygenase is generally inhibited at very low concentrations of these drugs, perhaps lower than the therapeutic tissue concentration. Aspirin irreversibly inhibits cyclooxygenase by acetylation of the active site, as shown in Figure 7.8. It is worth noting that the analgesic and antipyretic drug paracetamol is not an effective inhibitor of cyclooxygenase.

Box 7.5 considers the discovery of the inhibitory effects of aspirin on prostaglandin synthesis. The possibility that this finding could not only account for the chemical mediation of inflammation but could also explain the mechanism of action of the NSAIDs generated massive interest.

Given that prostaglandins released locally in the tissues are able to promote vasodilatation as well as increase vascular permeability and pain, they are

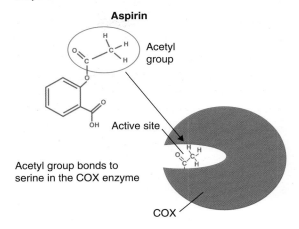

FIGURE 7.8 Aspirin inhibits cyclooxygenase (COX) by acetylating the active site, thus inactivating the enzyme

obvious candidates as mediators of inflammation. As they are also involved in pyrexia, it is clear to see how the anti-inflammatory, analgesic, and antipyretic activities of NSAIDs can be explained by their ability to inhibit prostaglandin synthesis. In fact, their most commonly encountered side effect, gastrotoxicity, can also be explained by this mechanism, as prostaglandins are cytoprotective to the gastric mucosa. The synthetic prostaglandin analogue misoprostol is used to protect the lining of the stomach in some patients sensitive to the gastrotoxic effects of NSAIDs.

BOX 7.5

Prostaglandins: worthy of two Nobel Prizes

Research in the field of prostaglandins has been a fertile arena for Nobel Laureates. Ulf S. von Euler is credited with the original discovery of prostaglandins in 1935, which he found in semen and believed to have been produced in the prostate gland (hence the name, prostaglandin). He was joint recipient of the Nobel Prize for Medicine in 1970 for work on **humoral transmitters**. In 1971, John Vane and his colleagues published a series of papers in *Nature* which described the inhibitory effect of aspirin on prostaglandin synthesis. He was joint recipient of the Nobel Prize for Medicine in 1982 for his work on prostaglandins and related substances.

There is sustained interest in the use of aspirin, generally in small doses (75 mg per day) as an antiplatelet drug. Millions of people worldwide have used or are now using a small daily dose of aspirin as a prophylactic measure to prevent an undesirable thrombotic event such as myocardial infarction or stroke. Aspirin is also indicated following an ischaemic attack and following coronary artery bypass surgery. Low doses have been shown to preferentially inhibit the cyclooxygenase enzyme in platelets (reducing TXA_2 synthesis) with lesser effects on the generation of PGI_2 by endothelial cells.

 Section 9.6 'Haemostasis', within Chapter 9 of this book, further explores platelet aggregation and antiplatelet therapies.

A more recent development in the field of cyclooxygenase inhibitors has been the introduction of drugs that are selective inhibitors of cyclooxygenase 2 (COX2). The concept behind the development of such drugs is that the proinflammatory prostaglandins are largely synthesized by the inducible COX2 enzyme whereas the cytoprotective prostaglandins required in the stomach are the product of the constitutively expressed COX1 enzyme. Selective inhibition is an obvious therapeutic target.

There is some evidence to suggest that selective COX2 inhibitors such as celecoxib and etoricoxib are effective anti-inflammatory agents and do provoke less gastrotoxicity. Unfortunately, COX2 inhibitors have been found to be associated with an increased incidence of thrombotic crises. Therefore it is now recommended that these selective inhibitors are only used in patients who are particularly at risk of developing peptic ulcers. COX2 is a significant source of PGI_2, the antithrombotic prostaglandin. Selective inhibition of this enzyme may adversely affect the delicate balance between PGI_2 and TXA_2, which stimulates platelet aggregation.

KEY POINT

Non-steroidal anti-inflammatory drugs are very competent inhibitors of cyclooxygenase. These drugs will inhibit the synthesis of prostaglandins and thromboxanes; this effect is seen at very low concentrations of the drug.

SELF CHECK 7.5

What is the theoretically ideal dose of aspirin that would prevent undesirable thrombus formation?

Leukotriene antagonists

The leukotriene receptor antagonists montelukast and zafirlukast are effective antagonists of the leukotrienes LTC_4, LTD_4, and LTE_4. They have been found to be effective in asthma and are generally used in conjunction with other drugs, notably inhaled corticosteroids.

7.6 Hormones

True hormones are chemical messengers released by endocrine glands and disseminated around the body by the bloodstream. Their targets may be particular cells or tissues or they may have a general impact on the whole body. Hormones allow a coordination of activity that is consistent throughout the body; however, the response to their release takes longer to be manifested than changes seen in response to nerve-transmitted messages. It generally takes about 90 s for blood to pass around the body.

Most hormones, such as growth hormone, oxytocin, and insulin, stimulate receptors on the surface of their target cells. The typical response to the stimulation of these surface receptors is a change in the intracellular level of a substance which governs the overall specialist activity of the cell. A common example of this is a modification of the level of the *second messenger* cAMP. The hormone receptor will be closely allied to the enzyme adenylate cyclase via a G-protein. Stimulation of the G-protein-coupled receptor by its activating neurotransmitter or hormone (first messenger) will lead to the activation of the enzyme and a consequent increase in the metabolism of adenosine-5′-triphosphate (ATP) to cAMP. The activity of the cell will rise in relation to this increase in cAMP levels. This type of arrangement is very common throughout the body. Other second messengers include *inositol 1,4,5-trisphosphate* (IP$_3$) and *diacylglycerol* (DAG), both generated from *phosphatidylinositol 4,5-bisphosphate* (PIP$_2$) and *3′,5′-cyclic guanosine monophosphate* (cGMP).

Steroid hormones such as oestrogen, testosterone, and corticosterone differ markedly from most hormones in their interaction with their target cells. Steroid receptors are found within the cells they interact with. The steroid interacts with the intracellular receptor and the combination is translocated to the nucleus, where it modifies gene transcription, leading to a change in the synthesis of some of the protein products of the cell. This, of course, can substantially change the activity of the cells.

 Section 4.2 'Basic concepts of pharmacodynamics', within Chapter 4 describes the different types of receptors in more detail while Section 2.3 'The instruction manual of life: the genetic code', within Chapter 2 details the process of gene transcription.

> **KEY POINT**
>
> Neural control of body function is rapid and precise. Hormonal control of function is slower but allows a consistent, coordinated response throughout the body.

The pituitary gland

One of the most significant arrangements for hormonal control within the body is the release of hormones that is indirectly controlled by the nervous system. The pituitary gland, shown in Figure 7.9, is found just under the base of the brain and attached to the hypothalamus by a blood-perfused stalk. It secretes several very significant hormones, which are listed in Table 7.2. These hormones are intimately and continuously involved with the normal, healthy running of the body. Their release is stimulated by

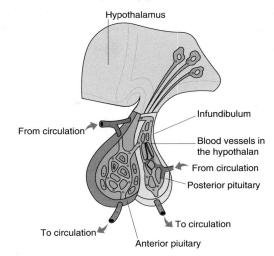

FIGURE 7.9   The pituitary gland

the action of a range of releasing factors delivered to the hormone secretory cells from the hypothalamus via the *hypothalamic–hypophysial portal system*. The releasing factors are generated by the hypothalamus in response to nervous activity.

The pituitary gland may be subdivided into regions. Oxytocin and vasopressin are released from the posterior section of the gland, melanocyte-stimulating hormone from the central region, and the other hormones from the anterior part of the pituitary gland.

Thyroid-stimulating hormone

Thyrotropin-releasing hormone (TRH) is released by the hypothalamus and stimulates the release of thyroid-stimulating hormone (TSH) from the anterior pituitary gland. TSH enters the circulation and is transported to its principal target, the thyroid gland. In the thyroid gland it stimulates TSH receptors on follicular cells to produce and release two very similar thyroid hormones, thyroxine and triiodothyronine, often referred to as T4 and T3, respectively. The most active thyroid hormone is T3. In the tissues, T4 is deiodinated to yield the more active T3.

Thyroid hormone levels in the circulation are controlled by an elegant negative feedback system. When high levels of T4 and T3 are detected in both the hypothalamus and anterior pituitary gland the release of TRH from the hypothalamus is reduced. This consequently reduces the release of TSH from the pituitary,

TABLE 7.2 Hormones secreted by the pituitary gland

Hormone	Target cells or organ and action
Adrenocorticotropic hormone (ACTH)	Release induced by corticotropic releasing hormone (CRH). Stimulates the adrenal cortex to release corticosteroids, which have a major role in defining which substrates are utilized to provide energy for metabolic processes.
Thyroid-stimulating hormone (TSH)	Release induced by thyrotropin-releasing hormone. Stimulates the thyroid gland to release the thyroid hormones T4 and T3, which stimulate the metabolic rate.
Growth hormone (GH)	Release induced by growth hormone-releasing hormone. It affects all tissues either directly or indirectly (by the release of other growth stimulants from the liver). Stimulates growth in children.
Prolactin	Release induced by prolactin-releasing factor. Stimulates mammary glands to produce milk.
Luteinizing hormone (LH) Known as interstitial cell-stimulating hormone (ICSH) in males	In females induces ovulation and promotes development of the corpus luteum. In males stimulates testosterone production.
Follicle-stimulating hormone (FSH)	Release induced by gonadotropin-releasing hormone. Stimulates the growth of follicles in females.
Melanocyte-stimulating hormones	Stimulate the production of melanin in the skin.
Oxytocin	Stimulates uterine contraction and milk secretion.
Antidiuretic hormone (ADH)	Stimulates the kidney to conserve water.

146

ultimately leading to the reduction of T3 and T4 levels in the blood.

Feedback control is very commonly found in biological systems. High levels of corticosteroids, for example, will inhibit the release of corticotropic releasing hormone from the hypothalamus, known as the *long feedback loop*. It will also inhibit further release of adrenocorticotropic hormone (ACTH) from the anterior pituitary gland, known as the *short feedback loop*.

KEY POINT

The pituitary gland acts as a mediator between the brain and other endocrine glands. Neural activity from the brain induces the secretion of hormones from the pituitary which stimulate the activities of peripheral endocrine glands.

Adrenal gland

The principal corticosteroid produced by the cortex (outer layer) of the adrenal gland in adult humans,

shown in figure 7.10, is *cortisol*. Cortisol is a *glucocorticoid* and is constantly present in blood. Normal low levels of cortisol are sustained by low-level stimulation of adrenocorticotropic hormone (ACTH) from the pituitary gland and feedback control.

FIGURE 7.10 The adrenal glands are found on top of the kidneys

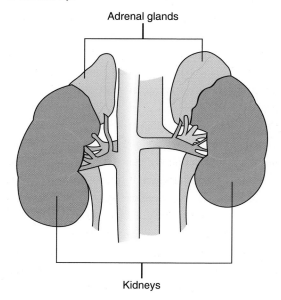

Adrenal glands

Kidneys

Higher levels of cortisol are released from the adrenal cortex in response to stress. Its primary effect is to increase blood sugar levels by promoting **gluconeogenesis**. As well as its function in the control of metabolic processes, cortisol has a large number of additional properties. It has, for example, a moderating effect on the body's inflammatory and immunological processes, a property that has significant therapeutic potential.

 Gluconeogenesis is covered in more detail in Section 3.3 'Metabolic pathways and abnormal metabolism', within Chapter 3 of this book.

Excessive secretion of cortisol gives rise to a complex disorder known as *Cushing's syndrome*. The most common form of the syndrome is known as Cushing's disease, which is caused by a tumour of the pituitary gland. This leads to the excessive secretion of ACTH, which in turn gives rise to excessive levels of cortisol. The most obvious signs of this syndrome relate to changes in the utilization of energy sources and the storage of preserved energy sources. The face becomes rounder and a build-up of fat stores between the shoulder blades gives rise to what is described as 'buffalo hump'. Muscle bulk is lost and the legs may become noticeably thinner.

Glucocorticoids

Glucocorticoids are very powerful inhibitors of the inflammatory process and are also potent immunosuppressants. Derivatives of these molecules, such as the drugs prednisolone and betamethasone, are often used to treat patients with serious acute manifestations of inflammatory disease. Patients treated with systemic glucocorticoids for more than a few days may develop changes very reminiscent of Cushing's syndrome. These changes are generally referred as being Cushing-like or cushingoid.

Systemic treatment with glucocorticoids is associated with another potentially serious issue. High levels of administered glucocorticoids will inhibit the release of corticotropin-releasing hormone (CRH) and consequently ACTH. As a result the adrenal cortex will not receive its normal constant secretory stimulation. Without routine stimulation with ACTH the adrenal cortex will regress and may lose its ability to synthesize and secrete normal levels of cortisol.

> **KEY POINT**
>
> Glucocorticoids have a significant responsibility to govern the way in which the body selects nutrients to metabolize for energy. They are also very powerful anti-inflammatory agents.

Pancreas

Another example of hormone control can be seen associated with the pancreas. The pancreas works largely independently of control from the pituitary gland but nevertheless fulfils a vital role. The pancreas is a diffuse gland found embedded in the **mesentery** adjacent to the **duodenum**, as shown in Figure 7.11. It is both an *endocrine gland* (one that secretes hormones into the circulation) and an *exocrine gland* (substances secreted via ducts). Its exocrine gland function is to synthesize and excrete digestive enzymes and release them via the pancreatic duct and the common bile duct directly into the duodenum, where they are able to act on the partially digested food material (chyme) present there.

The endocrine function of the pancreas is fundamental to the control of blood sugar levels in the body. Controlling the level of glucose in the body within very fine limits is essential. It is important that sufficient glucose is available to sustain normal body activities at any time; however, excessive levels of glucose are damaging. Over a period of time, high blood levels of glucose will result in the **glycation** of many normal body proteins, resulting in some loss of function.

A typical meal will contain a range of carbohydrates, which may include monosaccharides such as glucose, disaccharides such as sucrose or lactose, and the polysaccharide starch. Following ingestion, disaccharides and polysaccharides will be converted to monosaccharides, principally by the action of pancreatic **amylase** and disaccharidase enzymes released from the lining cells of the duodenum. Once rendered available, glucose is taken up into the blood by a co-transport mechanism driven by differential sodium ion levels, leading to an increase in blood glucose levels.

Glucose levels in the blood are detected by cells in the pancreatic **islets of Langerhans**. When glucose

FIGURE 7.11   The pancreas

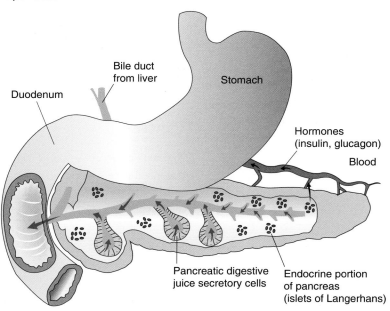

Bile duct
from liver

Stomach

Duodenum

Hormones
(insulin, glucagon)

Blood

Pancreatic digestive
juice secretory cells

Endocrine portion
of pancreas
(islets of Langerhans)

levels rise the hormone *insulin* is released by β-cells in the islets. Insulin stimulates the uptake of glucose by a large proportion of the body's cells, most notably in the liver, muscle, and adipose tissue. Glucose is converted to the polysaccharide *glycogen* in the liver and skeletal muscle for storage. The generation of glycogen (glycogenesis) enables the storage of glucose without significantly raising the concentration of glucose molecules in cells, which would otherwise impede subsequent uptake.

 Glucose metabolism is covered in Section 3.3 'Metabolic pathways and abnormal metabolism', within Chapter 3 of this book.

Some time after a meal, when excess glucose has been stored and the body has utilized the circulating glucose, the blood glucose level will fall. This fall will be enhanced by higher levels of activity such as exercise. This fall in blood glucose level is a significant factor in the development of sensations of hunger. It is also monitored by the pancreas, which will respond by releasing a different hormone. In these circumstances, the α-cells of the islets will release *glucagon*. The most obvious effect of glucagon is to stimulate the breakdown of glycogen to glucose in the liver (glycogenolysis) and the export of this glucose into the blood for circulation to the tissues of the body.

KEY POINT

The level of glucose in the blood is mainly controlled by the release of hormones from the pancreas. Insulin lowers blood glucose levels by promoting its uptake into the liver and skeletal muscles, while glucagon will raise blood levels by stimulating the release of glucose from the liver.

It is important to note that there are many other hormones, such as adrenaline (epinephrine), thyroxine, and corticosteroids, which can significantly influence blood glucose levels in certain circumstances. In general, however, it is the pancreas that has the principal task of maintaining glucose levels in most normal circumstances.

Diabetes mellitus

There are circumstances when blood glucose levels are not controlled adequately. Diabetes mellitus is a condition in which blood glucose levels are consistently higher than normal. It is estimated that between 2 and 3 million people in the UK have diabetes, though many of them are unaware of their condition.

There are two major classes of diabetes:

- Type 1 diabetes, also known as insulin-dependent diabetes, is usually initiated in young people. In

TABLE 7.3 Target blood glucose levels before and after a meal

Situation	Blood glucose level (millimoles per litre)
Normal—before a meal	3.5–5.5
Normal—2 hours after a meal	<8
Type 1 diabetic child—before a meal	4–8[†]
Type 1 diabetic child—2 hours after a meal	<10[†]
Type 1 diabetic adult—before a meal	4–7[†]
Type 1 diabetic adult—2 hours after a meal	<9[†]
Type 2 diabetic adult—before a meal	4–7[‡]
Type 2 diabetic adult—2 hours after a meal	<8.5[‡]

[†]NICE clinical guideline CG15. *Type 1 Diabetes: Diagnosis and Management of Type 1 Diabetes in Children, Young People and Adults.* National Institute for Health and Clinical Excellence, 2004.
[‡]NICE clinical guideline CG66. *Type 2 Diabetes: the Management of Type 2 Diabetes (Update).* National Institute for Health and Clinical Excellence, 2008.

this condition the pancreas is unable to produce sufficient insulin to promote controlled storage of glucose. The individual will generally self-administer insulin to overcome the deficiency. Table 7.3 shows target levels of glucose in a type 1 diabetic individual before and after a meal.

- Type 2 diabetes, also known as adult-onset diabetes, occurs later in life. It is the more common form of diabetes. In this condition the individual may produce even higher than normal amounts of insulin but the cells of the body do not respond adequately to it. Table 7.3 shows target levels of glucose in a type 2 diabetic individual before and after a meal. Mild forms may be treated mainly by restricting carbohydrates in the diet. Evidence is emerging to suggest that this approach may be able to reverse the pathology. In more severe cases orally active hypoglycaemic drugs such as

metformin (a biguanide) may be used. This drug inhibits gluconeogenesis and stimulates the utilization of available glucose. Type 2 diabetes may also be treated with a sulfonylurea such as glibenclamide, which enhances the secretion of insulin.

KEY POINT

It is very important for long-term health that glucose levels are maintained at very precise levels. There are two common ways in which the necessary control may be lost. In children the ability to produce insulin may be reduced, while in adults the ability to respond to insulin may be impaired.

SELF CHECK 7.6

What is diabetes and how can it be controlled?

CHAPTER SUMMARY

➤ Autocoids are responsible for the local, fine control of a range of body systems. Autocoids produce their effects by diffusion.

➤ Histamine has a major role in the normal digestive process and a very significant role in the pathology of acute allergic inflammation.

➤ Serotonin is involved in the normal function of platelets and has a role to play in normal peristalsis. It is also an important neurotransmitter in the brain.

➤ Kinins may have a role in functional dilation associated with sweat gland activity.

➤ Eicosanoids have a wide range of properties and may be involved in a large number of physiological and pathological processes.

➤ Prostaglandins have a range of properties and may be involved in the fine control of a wide range of different physiological processes. Their synthesis is dramatically inhibited by non-steroidal anti-inflammatory drugs.

➤ Hormones are responsible for the control of whole body systems. Their effects are achieved by dissemination throughout the body via the bloodstream.

➤ The brain induces the release of several hormones from the pituitary gland.

➤ Pituitary hormones stimulate the activity of many body systems, often by the release of other hormones. Adrenocorticotropic hormone, for example, induces the release of glucocorticoids from the adrenal cortex.

➤ Blood sugar levels are controlled principally by hormones of the pancreas. Errors in this control system can lead to diabetes.

FURTHER READING

Greene, R.J. and Harris, N.D. *Pathology and Therapeutics for Pharmacists: A Basis for Clinical Pharmacy Practice.* 3rd edn. Pharmaceutical Press, 2008.

This textbook is a very useful source of information on the rational use of drugs in clinical practice as it explores the use of drugs in the context of the underlying pathology of the diseases for which they are used.

Brunton, L., Chabner, B., and Knollman, B. *Goodman & Gilman's The Pharmacological Basis of Therapeutics.* 12th edn. McGraw-Hill, 2010.

This textbook offers probably the most authoritative and detailed review of the pharmacological activities of drugs currently available.

Homeostasis

PETER PENSON AND NEIL HENNEY

The human body can survive in a remarkable range of environments and situations. Perhaps you are reading this on a cold winter's day when there is snow on the ground, or maybe it is a scorching day in the summer. In either case your internal (core) body temperature is close to 37°C. *Thermoregulation*, the control of internal body temperature, is one example of *homeostasis—the body's control of its internal environment*. Homeostasis is a general term that refers to the many ways in which the body reacts to changes in its surroundings in order to maintain near-constant internal conditions. 'Homeostasis' is derived from two Greek words and could be translated as 'steady state' or 'standing still'. This does not mean that nothing in the body ever changes, but that life is sustained by maintaining important factors such as temperature and the concentration of oxygen and glucose in the blood within well-defined limits. This chapter will give several examples of homeostatic regulation in important biological systems to demonstrate the concept of homeostasis, but you should not think that homeostasis is limited to these examples. After reading this chapter you are encouraged to find examples elsewhere in this book of the body controlling its internal environment.

Learning objectives

Having read this chapter, you are expected to be able to:

➤ Give a definition of homeostasis.

➤ Explain the concept of negative feedback and describe its importance in homeostasis.

➤ Explain how homeostasis occurs in at least three body systems.

➤ Give examples of diseases or conditions which are caused by defective homeostasis and explain the mechanisms of disease in each case.

8.1 Regulating the internal environment

Homeostasis is often compared to the state of *dynamic equilibrium*, in which a reversible chemical reaction proceeds at the same rate in the forward and reverse directions and there is no net change in the concentration of reactants and products. However, this analogy can be misleading. Homeostasis is not about keeping the body in a fixed and unaltered state. It is about appropriately regulating the internal environment in response to the external environment (see Box 8.1). For example, when we exercise our muscles need more oxygen than

You can read more about the synaptic transmission of acetylcholine in Section 5.2 'An exciting journey: how are signals transmitted?', within Chapter 5, and about receptors for drugs in Section 4.2 'Basic concepts of pharmacodynamics', within Chapter 4.

BOX 8.1

Joseph Barcroft

Sir Joseph Barcroft (1872–1947) was a British physiologist who made important discoveries in the fields of **respiration** and thermoregulation—often using himself as an experimental subject. He once exposed his body to such extreme cold that he became unconscious. He also spent a week in a glass chamber in order to determine how much oxygen the human body needs in order to survive. His work improved our understanding of homeostasis in a number of systems in the body.

when we are at rest. So when we exercise, our heart rate and breathing rate increase in order to supply this increased oxygen demand.

The ability of the body to survive extremes of temperature, humidity, hunger, and thirst, and to endure exercise, is noteworthy given that individual components of the body are very susceptible to damage caused by changes in their surroundings. Almost all of the chemical reactions necessary for life depend on the action of **enzymes**. They are a type of biological catalyst and without their assistance many of the chemical reactions in the body would occur too slowly to sustain life. Have a look at Integration Box 8.1, which considers the importance of the enzyme acetylcholinesterase.

INTEGRATION BOX 8.1

Acetylcholinesterase

Effective transmission of neuronal signals requires that signals are initiated and terminated rapidly. We will briefly consider the process leading to contraction of skeletal muscle. An **action potential** in the **presynaptic neuron** leads rapidly to release of the neurotransmitter **acetylcholine** at the synapse by **exocytosis**. Acetylcholine binds with **nicotinic acetylcholine receptors** at the **motor endplate**. This leads to **depolarization**, which spreads across the muscle fibre and leads to contraction of the muscle. In order to maintain coordinated control of muscle, it is necessary to terminate the signal rapidly. This is achieved by the enzyme acetylcholinesterase, which breaks down acetylcholine in the **synaptic cleft**, thereby inactivating it.

The activity of acetylcholinesterase is dependent on temperature. Its optimal activity is around core body temperature, 37°C. Below this temperature the breakdown reaction will occur more slowly—as predicted by normal chemical kinetics. However, above normal temperature the heat **denatures** the protein molecule and activity falls. Between 42 and 48°C the activity of acetylcholinesterase falls by about 60%. This is typical of many enzymes and other proteins in the body, and protein denaturation explains why body temperatures above 40°C are often life-threatening.

A rise in body temperature of a few degrees can lead to catastrophic disruption of metabolic processes. The process by which the body maintains a near-constant temperature will be discussed in Section 8.3. Temperature regulation illustrates a very important point—that disruption of homeostasis has very severe effects on the body; indeed *almost all diseases* can be attributed to the *dysfunction* of a homeostatic mechanism.

KEY POINT

Disruption of homeostasis leads to disease states and death.

Not all organisms have the ability to regulate their internal environment in response to changes in the external environment to the same extent that we do. In general, complex organisms have very well-developed mechanisms of homeostasis and simple organisms have less well-developed means of controlling their internal environment. However, even single-celled organisms have evolved methods of regulating their internal environment to some extent. For example, amoebae (especially those which reside in fresh water) are liable continually to gain water from their surroundings by osmosis through the cell membrane. This has the effect of diluting the cell contents. Look at Figure 8.1, which shows how the amoeba opposes these changes by the use of a contractile vacuole—a

FIGURE 8.1 Osmotic regulation in an amoeba. (A) The osmolarity of the intracellular fluid is greater than that of the fluid surrounding the amoeba. Water enters the cell by osmosis through the cell membrane (indicated by the arrows), reducing the intracellular solute concentration. (B) Water moves from the cytoplasm into the contractile vacuole. (C) The vacuole fuses with the cell membrane and 'contracts' releasing its contents into the extracellular space

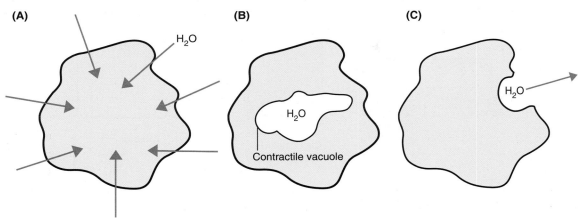

(A)
H_2O

(B)
H_2O
Contractile vacuole

(C)
H_2O

membrane-bound compartment within the cytoplasm (but very close to the plasma membrane) that fills with water and then fuses with the cell membrane, releasing its contents outside the cell and thereby restoring the fluid balance.

SELF CHECK 8.1

Give a definition of homeostasis. Why is homeostasis important?

153

8.2 Feedback mechanisms to achieve homeostasis

In order for the body to regulate its internal environment, a system with three elements is required:

1. A *sensor* that can measure a property of the internal environment. This could be an **ion channel** or some other molecule that changes its shape or activity in response to a change in its surroundings.

2. A *control centre* which processes the information from the sensor.

3. An *effector* that carries out an action in response to a signal from the control centre. This could be a muscle or **gland**, a specialized area of tissue, or indeed a whole organ.

In order to understand this it is helpful to consider the analogy of a very simple central heating system such as there might be in a house. Look at Figure 8.2, which shows how this system is set up with a thermostat and a heater.

The sensor (thermometer), control centre (circuit board), and effector (heater) work together to control the temperature inside the room. The switching on of the heater leads to an increase in the temperature in the room. This is detected by the thermometer and when the temperature surpasses that set, the control centre switches off the heater. When the products of the system (in this case heat) prevent their own production, the system is said to display *negative feedback*.

This type of negative feedback mechanism is very common in pathways of homeostasis in the body as it tends to produce stable systems. Negative feedback is important in keeping a factor such as core temperature

FIGURE 8.2 A schematic of a very basic heating system demonstrating the principle of negative feedback

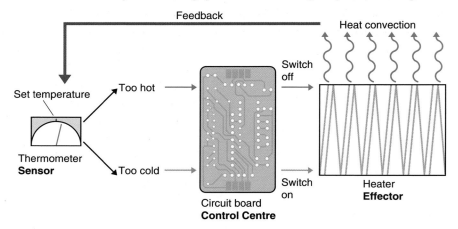

within well-defined limits. In the case of the central heating system, if the system did not display negative feedback the heater would stay on forever and the room would become far too hot. Because the increased temperature leads to the switching off of the heater, the room is kept at a comfortable temperature. The temperature is not a constant but is kept within defined limits. This is also true of many parameters within our bodies controlled by homeostatic mechanisms.

KEY POINT

Negative feedback describes the situation in which the products of a system reduce the activity of the system that produces them.

KEY POINT

Homeostasis generally relies on negative feedback mechanisms to keep biological parameters within well-defined limits.

Conversely, *positive feedback* refers to the situation where the products of a system increase the activity of the system that produces them. Unlike negative feedback, positive feedback does not tend to stabilize systems and therefore positive feedback is not often associated with homeostasis. Indeed, the pathophysiology of many diseases includes positive feedback mechanisms. For instance, in heart failure the heart is unable to pump blood at a rate sufficient to meet the body's needs. The oxygen tension in the blood falls and the sympathetic nervous system is activated in response. In healthy patients the sympathetic nervous system increases the force and rate of contraction of the heart in response to short-term increases in demand during exercise. However, in patients with heart failure the continual activation of the sympathetic nervous system increases the workload of the already struggling heart and accelerates its decline. Cardiac output is reduced further, which leads to an even greater activation of the sympathetic nervous system. This vicious cycle is very difficult to resolve.

 The effect of the sympathetic nervous system on the heart is detailed in Section 6.4 'Effect of the ANS on the cardiovascular system' within Chapter 6 of this book.

There are very few situations where positive feedback mechanisms are beneficial to the body, but they do exist. For example, when a blood vessel is injured and bleeding occurs, proteins such as collagen are exposed in the damaged region of blood vessel and platelets bind to the collagen, beginning the process of *haemostasis* (not to be confused with homeostasis) in which the combination of platelet plug formation and blood coagulation prevent further blood loss. This requires very many platelets to be recruited from the circulation to the damaged region of blood vessel. This is achieved by the fact that when platelets are recruited to the site of injury and activated they release adenosine-5′-triphosphate (ATP), which is detected by circulating platelets, which are then attracted to the site of injury. These newly arriving platelets release more ATP, amplifying the attractive signal and thus creating a positive feedback loop, which ensures that platelets are rapidly recruited and form a clot to stop the flow of blood.

 The process of haemostasis is explained in more detail in Section 9.6 'Haemostasis', within Chapter 9 of this book.

SELF CHECK 8.2

Explain what is meant by negative feedback and why it is important in homeostasis.

Redundancy in homeostasis

We can extend the analogy of the simple heating system to illustrate some other important features of homeostasis. Think of the way it can be used to regulate the temperature in the room. If it is too cold the central heating can be switched on. However, the room might still feel too cold, so the thermostat on the heater can be adjusted. The curtains can also be closed to reduce heat loss through the windows. If it is still not warm enough (or if the central heating is broken) another method of warming the house such as an electric heater or a log fire could be employed. By combining a number of these approaches we can ensure that the room is always at a comfortable temperature.

In general, our bodies also have numerous mechanisms that lead to the same outcome. Some of these will be described in this chapter. Having more than one mechanism to achieve the same outcome is known as *redundancy*. Redundancy is important because it allows fine control and also gives the body a backup mechanism if the major mechanism of control is ineffective owing to illness or injury.

KEY POINT

Homeostatic mechanisms are complex and display redundancy.

SELF CHECK 8.3

Explain what is meant by redundancy and why is it useful in homeostasis.

8.3 Major homeostatic mechanisms in the human body

All organs and tissues of the body are involved in the maintenance of an internal environment. Important examples include:

- Regions of the *brain* (particularly the hypothalamus) are the most common 'control centres' for homeostasis. They serve to receive sensory information and send appropriate neuronal or hormonal signals to effectors.

- The blood flow to *skeletal muscle* is continually adjusted according to demand. Shivering in skeletal muscle is an important way in which body temperature can be increased.

- Blood flow to the *skin* can be controlled so that blood is directed close to the surface of the skin or deeper internally. This is also important in thermoregulation.

- The rate and force of the contractions of the *heart*, and the rate and depth of *breathing*, are continually adjusted to maintain sufficient partial pressures of oxygen in the blood and organs.

- The *pancreas* releases glucagon and insulin, which are essential for blood sugar homeostasis.

- The *kidneys* have a number of homeostatic functions. They are essential for maintaining fluid balance.

To illustrate the general points made about homeostasis, several examples in human body systems will be given. These are not intended to be exhaustive—homeostasis, in different guises, occurs in all body systems. Nevertheless, it is hoped that these examples will illustrate the breadth and complexity of homeostatic systems.

Temperature regulation

As we saw in the introduction to this chapter, regulation of body temperature is very important to enable complex organisms to survive and function. Organisms such as mammals, which create most of their body heat as a result of metabolic processes and carefully regulate their own temperature, are referred to as *endotherms* or 'warm-blooded animals'. *Ectotherms* or 'cold-blooded animals' derive most of their heat from their surroundings (for example reptiles bask in the sun to regulate their body temperature). As a result, the body temperature of ectotherms is generally lower than that of endotherms and it is more prone to variation as the environmental temperature alters.

The temperature surrounding our internal organs (37°C) is usually several degrees higher than the temperature at the skin of our fingers and toes and is less prone to variation as the environmental temperature changes. When we talk about our body temperature, we are referring to our *core temperature*, the temperature deep inside the body. Refer to Box 8.2 for measurement of core body temperature.

SELF CHECK 8.4

Explain the difference in temperature regulation between ectotherms and endotherms.

BOX 8.2

Measuring core body temperature

When a doctor measures a patient's temperature they will place the thermometer under the patient's tongue, in their armpit, or in their ear, because this is as close as we can get to reliably measuring core temperature. Experimental studies have shown that the measurement of core temperature varies slightly according to the site used. Thus it is important to use a consistent technique when monitoring changes in a patient's body temperature at different stages over a period of time.

As we know, normal human core body temperature is around 37°C. A group of neurons in the hypothalamus acts as the sensor and control centre for body temperature in the same way that the thermostat does in the central heating system we have described. If the core temperature creeps above or drops below this normal value, the hypothalamus can initiate a series of neuronal and hormonal signals to specialized organs and tissues in order to initiate a negative feedback pathway that will stabilize the body temperature at the normal value.

If the body temperature is too high the following techniques are used to transfer heat away from the body:

- *Sweating*—stimulated by the activation of sympathetic neurons, which innervate the sweat glands found in the skin. Unusually for sympathetic neurotransmission, these postsynaptic neurons release acetylcholine rather than noradrenaline. Acetylcholine causes the sweat glands to release their contents onto the surface of the skin. Sweat consists of a watery solution of urea, chloride ions, and some other minerals. The water component of sweat evaporates from the body. This transfers heat away from the body because of the latent heat of evaporation of the water. The efficiency of sweating to reduce body temperature can be severely impaired if a person wears clothing that is impermeable to water and therefore prevents evaporation of sweat. It is important to realize that heat loss during sweating occurs at the expense of water loss. Therefore temperature homeostasis has an effect on fluid homeostasis. During hot weather it is important to drink more water than usual in order to replace that which is lost by sweat and thereby reduce the risk of dehydration.

- *Peripheral vasodilatation*—blood flow within the body is matched to the oxygen and nutrient demands of different organs and tissues. Blood flow distribution is regularly adjusted rather than being constant. For instance, after eating a meal, blood flow is directed towards the gastrointestinal tract to aid the absorption of nutrients, whereas during exercise, blood flow is directed to skeletal muscles. This control of blood flow is achieved by altering the diameter of the arterioles that supply the skin with blood. Arterioles are small blood vessels that can be found between arteries and capillaries. The walls of arterioles are relatively rich in smooth muscle, which has large numbers of α-adrenoceptors and is innervated by sympathetic nerves. When the sympathetic nerves release noradrenaline onto the adrenoceptors, calcium ion channels open in the muscle cell membrane, and calcium ions flow into the cells, activating the contractile machinery of the cells, leading to contraction. When this smooth muscle contracts, the lumen of the blood vessel is reduced in size and downstream blood flow is reduced. This is called *vasoconstriction*. *Vasodilatation* is the opposite process, whereby arteriolar smooth muscle relaxes and increases blood flow. The blood flow to the skin is carefully controlled in this manner. When the body is too hot, vasodilatation in the arterioles that supply the skin with blood occurs. This brings the warm blood closer to the cooler external environment and therefore aids transfer of heat away from the body.

- *Flattening of skin hair*—the skin of humans is covered with a layer of hairs. This is probably a vestigial reminder of our evolutionary ancestors, who were covered with much denser hair. Nevertheless, the relatively sparse hair on human skin can still serve to trap a layer of air close to the body, insulating it against heat loss. The thermal insulation provided by body hair is more effective when the hairs are standing on end rather than lying flat. The position of each hair is determined by a small muscle connected to the hair follicle called an *arrector pili*. When this muscle is contracted the hair stands on end and the arrector pili are visible as goose bumps on the skin. When the body is too hot these muscles relax, allowing the hairs to lie flat.

SELF CHECK 8.5

Explain the mechanism of the physiological changes that are initiated to reduce a high body temperature in humans.

When the body is too cold, the processes listed are reversed. Sweating is stopped, blood flow is diverted towards the core rather than the periphery, and arrector pili muscles contract and body hair stands on end. Additionally, the body can make use of heat generated by exothermic metabolic reactions in order to raise its temperature. This can be achieved by:

- Shivering, which involves the contraction of skeletal muscles. The respiratory rate in the muscles is increased during shivering and respiration generates heat.
- Non-shivering thermogenesis, whereby brown adipose tissue generates heat from the metabolism of fats. This source of heat generation is particularly important in newborn babies, who have poorly developed muscles and cannot shiver.

SELF CHECK 8.6

Explain the mechanism of the physiological changes in humans that are initiated to increase a low body temperature.

Look at Figure 8.3, which shows how thermoregulation consists of two negative feedback pathways that work in opposition to one another. This technique of employing opposing mechanisms helps to maintain stable systems and is therefore very useful in homeostasis. Integration Box 8.2 outlines the concept of *fever*, in which the core body temperature is increased.

KEY POINT

Homeostatic systems are often kept in balance by two separate negative feedback pathways working in opposition to one another.

Failure of temperature homeostasis: hypothermia and hyperthermia

Fever tends to maintain the body at a raised but fairly stable temperature (37–39°C), which returns to normal after the cause has passed. However, some illnesses can cause uncontrolled *hyperthermia* (high body temperature), which is much more dangerous because of the damage that can be done to biological molecules by high temperature. Hyperthermia can be caused by heat stroke—when the thermoregulatory mechanisms we have described are no longer adequate to reduce body temperature. Some antidepressant drugs that increase the level of serotonin in the brain can cause hyperthermia as an adverse effect. Hyperthermia is also an unwanted effect of several recreational drugs such as cocaine and amphetamine.

FIGURE 8.3 Negative feedback pathways in thermoregulation. The diagram shows the series of events that occur when the core temperature varies from the normal set-point

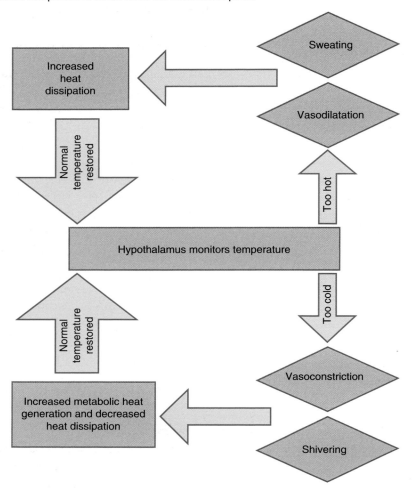

Seizures, unconsciousness, and death are the end results of prolonged severe hyperthermia.

 For a recap on serotonin see Section 7.3 'Serotonin', within Chapter 7; drug misuse is addressed in Section 4.4 'Drug misuse and addiction', within Chapter 4.

An abnormally low body temperature (*hypothermia*) usually occurs as the result of exposure to extremely cold environmental conditions. Body heat is generated by metabolic reactions, and the rate of these reactions drops when temperature falls. This can result in a positive feedback mechanism that takes the body temperature even further away from normal. In general the body is adapted to tolerate hypothermia better than hyperthermia. Because the metabolic rate is so low during hypothermia the oxygen demand of the tissue is very low, which can serve to protect the tis-

sue from damage. Nevertheless, sustained hypothermia leads to coma and death. Research has shown that the chance of surviving a cardiac arrest is increased if body temperature is kept low when the heart is not beating. Doctors sometimes therefore artificially reduce the body temperature of patients with cardiac arrest—this is called *therapeutic hypothermia*.

 You can read more about metabolism in Section 3.3 'Metabolic pathways and abnormal metabolism', within Chapter 3 of this book.

Blood pressure homeostasis

Regulation of blood pressure (BP) is another important example of homeostasis. BP is the force exerted by the blood on the walls of the blood vessels and is generated by the muscular contractions of the beating heart.

Fever

Body temperature is often raised in response to a bacterial or viral infection. This increased core temperature is referred to as fever. Fever results from a 'resetting' of the control centre in the hypothalamus so that it continues to regulate body temperature but keeps it at a higher temperature than normal. This occurs because the body recognizes the presence of *pyrogens* (substances which cause the temperature of the body to increase, such as bacterial cell wall lipopolysaccharides) and responds by releasing prostaglandin E_2. This eicosanoid is found in elevated concentrations in the **cerebrospinal fluid** during infections and is thought to mediate the effect whereby the hypothalamus is reset to keep temperature at a higher level than normal.

 Prostaglandins and eicosanoids are covered in more detail in Section 7.5 'Eicosanoids', within Chapter 7 of this book.

It is thought that the increased temperature may be beneficial to the body during infections because the rate of some immune reactions is increased. Furthermore it is argued that some microorganisms are more susceptible to changes in temperature than the human body and that although a small, short-lived rise in the temperature is not harmful to us, it may damage bacteria.

Despite the fact that fever is possibly beneficial in fighting an infection, doctors and pharmacists often want to reduce the patient's temperature. This is because fever is uncomfortable and the high body temperatures during an infection can lead to seizures, especially in young children. One way in which body temperature can be reduced is by the use of temperature-lowering drugs known as *antipyretics*. In addition to its **analgesic** effects, paracetamol is an antipyretic drug. It acts by preventing the synthesis of prostaglandin E_2, thereby abolishing the pyrogenic signal to the hypothalamus and restoring normal temperature.

It is this pressure that drives the blood into all organs and tissues to which it delivers oxygen and nutrients. The normal pressure of blood in the vessels varies throughout the body; for instance, BP in the *pulmonary circulation* (supplied by the right side of the heart) is much lower than the BP in the *systemic circulation* (supplied by the more muscular left side of the heart). Furthermore, because of the action of gravity upon circulating blood, the pressure is lower in the brain (above the level of the heart) than it is in the feet (below the level of the heart) when a person is standing upright.

BP changes continually throughout the cardiac cycle and is highest during cardiac contraction (*systole*) and is lowest during cardiac relaxation (*diastole*). By convention, the systemic arterial BP at the level of the heart is usually stated in terms of systolic/diastolic. Measurement is made with reference to the pressure produced in a column of mercury of known height. Thus a typical BP in a young person might be 120/80 mmHg (millimetres of mercury). This is not a standard Système International (SI) unit but refers back to when doctors used mercury manometers when measuring BP.

Many physiological factors control BP; hence there is significant redundancy in the control of BP. However,

BP can be said to be the product of cardiac output (the rate at which the heart pumps blood) and peripheral resistance (the resistance to blood flow in the blood vessels). Anything that alters BP does so by affecting one or both of these factors. Cardiac output is determined by the force and rate at which the heart contracts, and also by the blood volume. Thus a drug which reduces the force and rate of the contractions of the heart can reduce BP. Peripheral resistance is increased by *vasoconstrictors* such as noradrenaline, which cause contraction of smooth muscle in arterioles.

BP homeostasis is vital to life. If BP is too low, organs and tissues will not receive sufficient blood flow to meet their metabolic demands. Have a look at Box 8.3 for a dramatic example of what can occur if BP falls too low (*hypotension*). High BP is necessary during exercise to ensure that the skeletal muscle receives sufficient oxygen and nutrients, but prolonged high BP (*hypertension*) causes damage to all organs and tissues. The most damaging effects of high BP occur as a result of damage to a number of organs including: the heart, which can lead to myocardial infarction and heart failure; the kidney, leading to renal failure; and the brain, where stroke is a consequence of hypertension.

Fainting

It is easy to think of homeostasis in terms of the body slowly, gently, and silently making small adjustments to a number of factors simultaneously in order that certain important parameters in the internal environment are kept within normal limits. Thus we think of homeostasis as something that goes on in the background without us being aware of it. In many cases and situations this is true. However, in some cases the body reacts immediately and forcefully to life-threatening changes and we are immediately aware of the results.

Normally, if the brain detects that it is not receiving sufficient oxygen in the blood to meet its metabolic demands it causes an increase in heart rate to increase cardiac output to correct this problem. However, sometimes this reaction may not occur sufficiently quickly and the brain is at risk of damage. This low oxygen state provokes a number of symptoms, such as dizziness, pale skin, and weakness in the limbs. If the legs become weak the person may feel the need to sit or lie down or they may even faint (the medical term for fainting is *syncope*). When a person faints and falls to the floor, the BP in the brain will rise because the heart and brain are now at the same level and the heart does not have to work against gravity to pump blood into the brain. This is usually enough to restore a safe level of blood flow and oxygen to the brain.

Blood pressure (BP) constantly changes in order to meet the metabolic demands of the body. However, many factors control BP and this system also displays redundancy.

When we exercise, the increased respiratory activity in our muscles means that the heart must pump with greater force and at a greater rate to meet this elevated oxygen demand. This is achieved via the sympathetic nervous system and the hormone adrenaline, which is released from the adrenal medulla. Adrenaline can be released very rapidly in times of stress in order to prepare the body for action by increasing cardiac output.

The *renin–angiotensin–aldosterone pathway*, shown in Figure 8.4, is another important element in the homeostasis of BP. When BP in the renal artery falls, a proteolytic enzyme called *renin* is released from the juxtaglomerular cells in the kidney into the blood. This starts a complex series of events that serves as an effector to restore normal BP. Renin catalyses the breakdown of *angiotensinogen*, a protein which is produced by the liver and circulates in the blood. One of the breakdown products of angiotensinogen is a decapeptide (a protein consisting of ten amino acids) called *angiotensin I*. Although angiotensin I has no biological activity by itself, it can be further cleaved by *angiotensin-converting enzyme* (ACE), an enzyme found throughout the body but in particularly high concentrations in the lungs. Cleavage of angiotensin I by ACE leaves an octapeptide (a peptide with eight amino acids) called *angiotensin II*. Angiotensin II acts on receptors in the blood vessels to cause vasoconstriction. It therefore increases peripheral resistance. Angiotensin II also causes the release of a hormone, *aldosterone*, from the adrenal cortex. Aldosterone acts on the kidney tubule to increase the reabsorption of sodium and water, thereby expanding blood volume and increasing cardiac output. The combined physiological effects of angiotensin II and aldosterone lead to an increase in BP. This causes greater perfusion of the renal artery and therefore renin release is stopped. This negative feedback pathway is very effective in increasing BP because it raises both cardiac output and peripheral resistance.

The increases in BP that occur as a result of activation of the renin–angiotensin–aldosterone pathway are associated with the control of BP over longer periods of time, whereas rapid changes of BP are associated with activation of the sympathetic nervous system.

Failure of blood pressure homeostasis: hypertension

Unlike the example in Box 8.3, where a rapid fall in BP causes a faint, hypertension develops much more slowly. It is normal in developed countries to see an increase of BP with age. An average BP for someone in their late teens is around 120/80 mmHg and

FIGURE 8.4 The renin–angiotensin–aldosterone pathway.

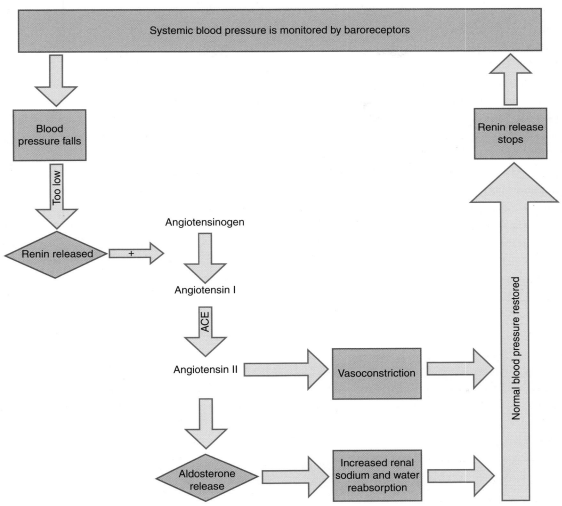

TABLE 8.1 Common antihypertensive drugs

Drug target	Drug class	Mode of action	Drug example
Renin	Renin inhibitors	Inhibition of the enzymatic activity of renin prevents the conversion of angiotensinogen to angiotensin I	Aliskiren
	β-adrenoceptor antagonists (β-blockers)	Renin release is promoted by the activation of β-adrenoceptors in the kidney; administration of β-blockers therefore reduces renin release	Atenolol
Angiotensin-converting enzyme (ACE)	ACE inhibitors	Inhibition of the enzymatic activity of ACE prevents the conversion of angiotensin I to angiotensin II	Captopril
Angiotensin II	Angiotensin II antagonists	The antagonism of angiotensin II receptors prevents vasoconstriction and aldosterone release	Losartan
Aldosterone	Aldosterone antagonists	Antagonism of the receptors for aldosterone prevents sodium retention and potassium secretion in the kidney thereby reducing blood volume; the vasoconstriction caused by angiotensin II is unaffected	Spironolactone

someone in their early sixties might be expected to have a BP of 135/90 mmHg, although there is a great deal of variation between individuals. In some tribal populations BP is remarkably invariant throughout adult life. Hypertension occurs when the BP is consistently raised to a level that is damaging to health. The choice of the upper limit of normal BP is entirely arbitrary, but most international guidelines for the treatment of hypertension aim to keep BP below 140/90 mmHg. You are probably asking yourself why BP rises as a person gets older—if so, you are in good company. Many scientists and doctors have tried to find the answer to this puzzle, although it would seem that there is no single cause. Hypertension occurs as the result of a failure of BP homeostasis but because many factors control BP and any one of them could go wrong it is very difficult to identify a universal cause. However, many people with high BP have abnormally high concentrations of renin circulating

in their blood, suggesting that there has been some disruption to the negative feedback process in the renin–angiotensin pathway.

Some drugs used to treat hypertension (*antihypertensive drugs*) target the renin–angiotensin pathway to reduce its efficiency at raising BP in order to try to restore normal homeostasis. Have a look at Table 8.1, which shows drugs that will commonly be found in community and hospital pharmacies and which act on the renin–angiotensin–aldosterone pathway to restore BP homeostasis. Case Study 8.1 considers how treatment with these drugs may be initiated in practice.

KEY POINT

Many antihypertensive drugs aim to restore homeostasis in the renin–angiotensin pathway.

CASE STUDY 8.1

Simon went for a regular check-up at his GP and was told he had high BP. He was advised to have another check-up the following week and was told he may have to take an ACE inhibitor if it remained high.

REFLECTION QUESTIONS

1. Why was Simon asked to come back for a check-up before any medication was prescribed?

2. What are ACE inhibitors and what consequences might you expect if Simon was given the wrong dose of his ACE inhibitor?

3. Simon had no idea that he had high blood pressure. Hypertension is usually symptomless. What consequences do you think this fact has for the treatment of hypertension?

Answers

1 Temporary increases in blood pressure occur as part of the normal homeostatic responses of the body. For example, when we are exercising, we need a higher blood pressure in order to supply blood to our rapidly working muscles. This is different to hypertension, a disorder in which disrupted homeostasis leads to a continuously raised blood pressure. Taking several measurements of blood pressure over a period of time will enable the doctor to establish whether Simon's BP is consistently high. It is important to treat high BP as it increases the risk of myocardial infarction (heart attack) and stroke. However, medicines may not be necessary to reduce his blood pressure. Changes in lifestyle, such as reduction of alcohol and salt intake as well as an increase in the frequency of exercise can reduce blood pressure too.

2 ACE inhibitors, such as captopril, inhibit an enzyme called Angiotensin Converting Enzyme or ACE. This enzyme is involved in the normal control of blood pressure. It is part of a system which increases blood pressure when it gets too low. In some people, the system is overactive and this leads to high blood pressure. ACE inhibitors restore normal blood pressure by blocking the actions of ACE. If the dose given were too high, he might experience hypotension and would perhaps feel dizzy. If the dose given were too low, his hypertension would not be adequately controlled and he would still be at increased risk of myocardial infarction and stroke.

3 Hypertension is often poorly controlled. This is because it is often undiagnosed unless the patient goes to see the doctor about an unrelated illness and the doctor measures their blood pressure. Furthermore, antihypertensive drugs, in common with all other medicines, exhibit side effects. Patients are often reluctant to take a medicine that gives them unpleasant symptoms when they are feeling otherwise healthy.

Blood glucose homeostasis

The monosaccharide glucose is the predominant source of energy for most cells under normal circumstances. Glucose can be metabolized during both aerobic and anaerobic respiration to produce the high-energy phosphate ATP, which is used ubiquitously in cellular reactions. Glucose is obtained from the diet directly and also from the breakdown of ingested complex carbohydrates (long chains of covalently linked sugar molecules) into their constituent sugars. Glucose is carried in the blood to all organs and tissues where it is utilized.

The normal concentration of glucose in the blood, the *blood sugar level*, is around 4–7 mmol/l although this varies somewhat throughout the day and is usually at its lowest first thing in the morning before breakfast. *Hypoglycaemia* is the term given to a low blood glucose concentration. The symptoms are unpleasant and include hunger, tachycardia, sweating, and irritability. Coordination can be lost and severe hypoglycaemia is often confused for drunkenness. Prolonged and extreme hypoglycaemia is very dangerous and results in seizures, coma, and eventually death.

Raised blood glucose concentration (*hyperglycaemia*) may initially be asymptomatic. However, prolonged hyperglycaemia has numerous detrimental effects to health, including damage to the eye (retinopathy), kidney (nephropathy), and peripheral blood vessels. Much of this damage is caused by *glycation*, which is the chemical bonding of glucose to proteins and other biological molecules. Furthermore, severe hyperglycaemia can result in a condition called *diabetic ketoacidosis*, which is a medical emergency with high mortality.

Given the severe consequences of hypoglycaemia and hyperglycaemia it is unsurprising that the body has evolved mechanisms of controlling blood glucose concentration. The negative feedback pathways involved in blood glucose homeostasis are shown in Figure 8.5.

Cells in the islets of Langerhans in the pancreas continually monitor the concentration of glucose in the blood. When the concentration rises (after a meal) the hormone *insulin* is released from specialized pancreatic cells called β-*cells*. Insulin promotes both the uptake of glucose into cells and the process of *glycogenesis*—the storage of glucose as glycogen. When blood sugar falls, the hormone *glucagon* is released from α-*cells* within the pancreas. Glucagon has opposing effects to insulin; it causes *glycogenolysis*—the breakdown of glycogen molecules into glucose.

The relatively simple scheme in Figure 8.5 does not show the whole story. A number of other hormones contribute to blood glucose regulation, including thyroxine, cortisol, and catecholamines (adrenaline and noradrenaline). Catecholamines reduce insulin release and promote glycogenolysis, thereby maintaining blood glucose concentrations during periods of stress.

Failure of blood glucose homeostasis: diabetes mellitus

Disruption of blood sugar homeostasis is the cause of diabetes mellitus. This condition can be caused by inadequate insulin release from the pancreas (type 1 diabetes) or by the inability of tissues to take up glucose in response to insulin (type 2 diabetes). In either case the result is hyperglycemia.

Under normal circumstances the kidney reabsorbs all the glucose that it filters. Therefore, no glucose is found in the urine of healthy people. However, in people with diabetes the blood glucose concentration can easily exceed the maximum capacity of the kidney to reabsorb glucose. This occurs when the concentration of glucose in the blood is around 10 mmol/l. Above this concentration glucose is lost in the urine. The glucose has an osmotic effect and therefore increases the production of urine by preventing reabsorption of water. Thus the three classic symptoms of diabetes mellitus are *glycosuria* (glucose in the urine), *polyuria* (increased production of urine), and *polydipsia* (increased thirst in order to compensate for the

FIGURE 8.5   Blood glucose homeostasis regulation by insulin and glucagon

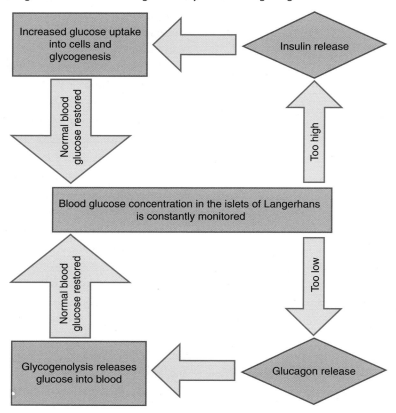

increased loss of water in the urine). Treatment of diabetes with insulin is discussed in Integration Box 8.3.

KEY POINT

Failure of blood glucose homeostasis leads to the clinical condition of diabetes mellitus.

SELF CHECK 8.8

Describe the consequences of hypoglycaemia and hyperglycaemia.

Calcium homeostasis

As well as the four common elements that make up the organic molecules in our bodies (carbon, oxygen, nitrogen, and hydrogen), we also require a much greater variety of minerals, such as magnesium, to keep us ticking over. Many of these minerals are required in only tiny quantities and we get them from eating a balanced diet. However, scientists are still not quite sure why we need them. However, it is clear that if we are deficient in them we may become sick in some way. For other minerals such as sodium, calcium, phosphate, potassium, iron, and many more, we do understand their uses much better, and we also know that deficiency or excess of these also leads to illness, sometimes to fatality in the extremes.

For example, our cells carefully regulate their internal levels of sodium and potassium (as well as other minerals) using proteins in the cell membrane which facilitate active transport (pumps) and pores (channels) which can be opened and closed to control this. Pumps require energy from ATP to work, but the cells must expend this otherwise they will not survive or function correctly. Control of sodium and potassium concentration is notably important for the functioning of nerve and muscle cells, and variation outside of narrow limits can lead to serious consequences such as cardiac arrest. As an example, we will consider one element that requires especially careful control in our bodies—calcium.

INTEGRATION BOX 8.3

Treatment of diabetes mellitus with insulin

The symptoms of type 1 diabetes mellitus can be controlled by the therapeutic administration of insulin. In the past, patients received porcine or bovine insulin, but this was not ideal because the peptide sequence of insulin from pigs and cows is not identical to human insulin and some people developed an immune reaction to the foreign protein. Nowadays we can give **recombinant human insulin** made by genetically modified bacteria. Because insulin is a peptide it cannot be given orally because it would be broken down by the **proteases** in the digestive system that convert the protein we eat into amino acids that can be absorbed. To bypass the digestive system insulin is delivered by **subcutaneous** injection. This is uncomfortable and inconvenient, especially for small children. A great deal of pharmaceutical research has revolved around the attempt to find novel drug delivery techniques for insulin. This has included exploration of **transdermal**, **intranasal**, and inhaled routes. This is an exciting area of ongoing research in the pharmaceutical industry.

 The different formations possible are outlined in Chapter 1 of the **Pharmaceutics** book within this series.

Calcium is a remarkable mineral. Not only is it an important part of our bone structure, it is also an important molecule in cellular signalling (so important in fact that calcium is often called the *universal second messenger*) and therefore control of cytoplasmic calcium concentrations is vital. Box 8.4 discusses the serendipitous discovery of the role of calcium in muscle contraction. Cells use intracellular calcium stores and calcium ion channels and pumps in the cell membrane to ensure that the cytoplasmic calcium concentration is appropriate for each signal or resting level. It is also important that extracellular calcium concentrations are carefully controlled. Look at Figure 8.6, which shows how this is achieved.

Calcium enters the body via the diet; it can leave the body via the urine and faeces or it can be stored in bone. The extracellular calcium concentration is mon-itored by the **parathyroid gland**. Calcium homeostasis is achieved predominantly through the use of three hormones:

- *Calcitonin* is released from the parafollicular cells in the thyroid gland in response to high plasma concentrations of calcium. It inhibits bone reabsorbing activity, thereby preventing calcium release from bone. Additionally it promotes calcium secretion by the kidney.

- *Parathyroid hormone* is released from the parathyroid gland in response to low plasma calcium. It increases reabsorption of calcium by the kidneys and increases bone reabsorbing activity, leading to the release of calcium from bone.

- *1,25-dihydroxycholecalciferol* (1,25-DHC) is the active metabolite of vitamin D_3. Parathyroid hormone also increases the production of this hormone, which maximizes calcium uptake from the intestines.

BOX 8.4

The importance of calcium

The discovery that calcium is required for muscle contraction was made by the British physiologist Sydney Ringer (1836–1910). In work published in the *Journal of Physiology* in 1882 he demonstrated that an excised frog heart could be kept beating by perfusing its blood vessels with what he thought was a saline solution (a solution of sodium chloride in water). However, he was unable to repeat his observations on a later occasion and set out to try to discover the reason. He discovered that when the experiments had been carried out for the published study, London tap water had been used instead of distilled water to make the saline. Analysis of the tap water showed that it contained calcium. Ringer then added calcium to the saline solution made from distilled water, and was able to replicate the previous results. The solution used nowadays to bathe isolated organs is known as Ringer's solution. The combination of a simple mistake in the laboratory and a sharp mind led to a very important discovery. Ringer was made a Fellow of the Royal Society in 1885; unfortunately history does not record the fate of the laboratory assistant who made the mistake.

FIGURE 8.6 Hormones involved in calcium homeostasis

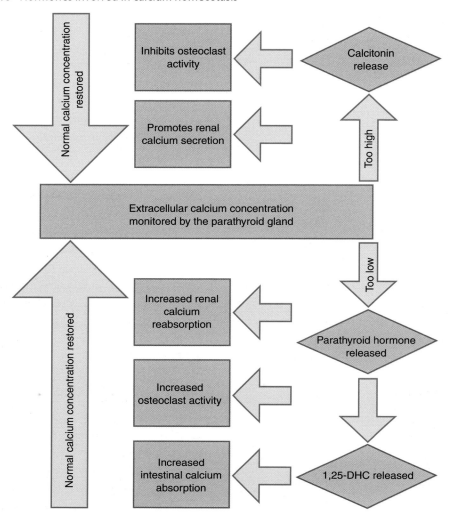

Calcium homeostasis in bones

It is tempting to think of the bones in our skeleton as fixed, unchanging—perhaps even dead. This is not true, as bone is a living tissue and like most other tissues it undergoes a continuous process of renewal. If this were not the case, broken limbs could not be repaired and growth would not be possible.

There are three predominant cell types in bone: osteoblasts, osteoclasts, and fibroblasts. These coexist with an organic matrix, which is predominantly made of collagen, and a mineral matrix of calcium phosphate crystals. Osteoblasts are 'bone forming' cells—they deposit calcium phosphate within the collagen matrix. Osteoclasts have the opposing effect—they

are 'bone reabsorbing' cells. These two processes must be kept in balance to ensure the structure of the bones remains intact. This balance between bone reabsorption and bone forming is an example of homeostasis. Parathyroid hormone and calcitonin help to regulate the activity of these cells, and work to increase uptake or release of calcium and phosphate in bones according to the levels and requirements elsewhere in the body. So the bones are not only structural but also a repository for essential minerals.

Osteoporosis, literally meaning porous bones, is a disease of the bones characterized by a low bone density and increased propensity to fracture. It is a significant cause of morbidity and is most prevalent in postmenopausal women. Osteoporosis occurs when the balance between osteoblast and osteoclast activity is lost, leading to insufficient bone deposition. Thus osteoporosis can be thought of as a disruption of homeostasis.

Name and briefly describe the actions of the hormones involved in calcium metabolism.

8.4 The kidneys: key organs in homeostasis

Examples of homeostasis can be found in all organs and tissues of the body; however, many mechanisms of homeostasis involve the kidney at some stage. This is not surprising given the function of the kidney as a 'gatekeeper' between the internal and external environments. In this chapter we have seen how the kidney is involved in regulating blood volume (and therefore BP), the regulation of blood calcium concentration, and the renal reabsorption of glucose. The kidney has many other functions that fall under the banner of homeostasis, including maintenance of body sodium and potassium levels, and maintenance of blood pH.

The kidney is an excellent example of all the features of homeostasis. It displays *negative feedback* pathways (such as the renin–angiotensin–aldosterone pathway described under 'Blood pressure homeostasis' in Section 8.3). It displays *redundancy* because blood volume can also be controlled by other renal mechanisms (such as **natriuretic peptides** and the hormone **vasopressin**). Finally, the kidney is an excellent example of

the way in which many complex homeostatic mechanisms operate in the same organ at the same time and yet manage to maintain a stable internal environment. The book *The Wisdom of the Body* by Walter Cannon, which is introduced in Box 8.5, first outlined the principles of homeostasis. The more we learn about the complexity of physiological mechanisms the more appropriate his title seems to become.

> ### BOX 8.5
>
> #### The wisdom of the body
>
> The concept of homeostasis resulted from the independent work of two scientists: the French physiologist Claude Bernard (1813–1878) and the American Walter Cannon (1871–1945). Cannon published a number of scientific papers giving his ideas on the subject, culminating with the publication in 1932 of a book called *The Wisdom of the Body*. This is an excellent account of the topic, written in a very clear and understandable way, and is well worth reading.

167

CHAPTER SUMMARY

➤ Complex organisms need to maintain a near-constant internal environment in order to sustain life.

➤ Homeostasis is the general term given to the processes by which the body maintains its internal environment.

➤ Most mechanisms of homeostasis rely on the principle of negative feedback in order to maintain stable systems.

➤ Stable systems require the continual careful adjustment of many variables within defined limits.

➤ Redundancy is the control of one physiological factor by a number of different mechanisms and is common in homeostasis.

➤ Disruption of the mechanisms of homeostasis leads to disease states.

Levick, J.R. *An Introduction to Cardiovascular Physiology.* 5th edn. Hodder Arnold, 2009.

> This book gives the physiological background to blood pressure, which is useful in understanding hypertension.

Pocock, G. and Richards, C. *The Human Body: An Introduction for the Biomedical and Health Sciences.* Oxford University Press, 2009.

> This book contains useful information on human anatomy and physiology and it goes into greater detail about calcium homeostasis, renal homeostasis, and glucose homeostasis than we have covered in this chapter.

Rang, H.P., Dale, M.M., Ritter, J.M., Flower, R., and Henderson, G. *Rang and Dale's Pharmacology.* Churchill Livingstone, 2011.

> This gives an excellent account of the therapeutic strategies for overcoming hypertension and other homeostatic imbalances discussed in this chapter.

Haematology

GILLIAN L. ALLISON

Haematology is a branch of medicine that is involved with the study of blood, blood-forming organs, and blood diseases, their prevention and treatment. It is important that we understand blood and its components, as it is the primary transport medium for nutrients and therapeutic molecules around the body.

In this chapter the major underlying theme is the functional role of blood cells in maintaining life. The different types of blood cell and their formation are discussed in this chapter. Focus is given to blood disorders such as anaemia and leukaemia, which target red blood cells and white blood cells, respectively. In addition, the process of preventing bleeding is explained and some of the antiplatelet, anticoagulant, and thrombolytic therapeutics available are mentioned. The ABO and rhesus blood grouping systems are also introduced with their implications on blood transfusions compatibility emphasized.

Learning objectives

Having read this chapter you are expected to be able to:

➤ Define the term haematology.

➤ Explain the process of blood cell formation, haematopoiesis, for both red blood cells (erythropoiesis) and white blood cells (leukopoiesis).

➤ Identify the components of blood plasma.

➤ Identify the major causes of the blood disorders anaemia and leukaemia, and explain the available therapeutic treatments for each disorder.

➤ Discuss the role of platelets as regulators of haemostasis, and describe the antiplatelet treatments that are currently available.

9.1 Introduction to haematology

Blood is a specialized **connective tissue** composed of three main cellular components: *erythrocytes* (red blood cells, RBCs), *leukocytes* (white blood cells, WBCs) and *platelets*. All these cell types are suspended in a complex non-living liquid matrix called *plasma*. The cellular components of blood are shown in Table 9.1.

Blood is a slightly basic (pH 7.35–7.45) liquid, with a greater density and viscosity than water, owing to the presence of the blood cells. It has three main functions: *transport*, *regulation*, and *protection*. It acts as a transportation medium for substances such as oxygen (O_2), carbon dioxide (CO_2), waste products, nutrients, hormones, and therapeutic agents to and from the organ

TABLE 9.1 The cellular components of blood

Blood cell	Size (µm)	Number (cells/litre)	Primary function
Erythrocyte, red blood cell (RBC)	7–8	$3.9–5.7 \times 10^{12}$	Transportation of oxygen
Leucocyte, white blood cell (WBC)	9–12	$4–11 \times 10^{9}$	Defence against infection
Platelet	2–4	$1.5– 4.4 \times 10^{11}$	Homeostasis

BOX 9.1

Appearance of blood

The red colour of blood is due to the presence of a red pigment called *haem*, which is part of the protein *haemoglobin* (Hb). The 'redness' of blood is dependent upon the amount of O_2 that is bound to the iron-containing haem molecule. For example, blood in arteries is bright red because it is highly oxygenated. This is in contrast to blood in the veins, which has a darker shade because there is little or no O_2 bound to the haem group. It should be noted that blood found in veins is not blue and that this concept is a myth. The blue colour is because the skin absorbs blue light much better than red.

systems and tissues of our bodies. Box 9.1 explains the effect of O_2 on blood colour. Blood is also important for maintaining homeostasis, by regulating body temperature, pH, and fluid volume within the body's tissues. Its protective role in the body is to prevent excessive blood loss through clotting mechanisms. Blood also helps in the protection against infection via the actions of the immune system.

You can read more about homeostasis in **Chapter 8** of this book.

FIGURE 9.1 The stages of blood cell development in the bone marrow (haematopoiesis). Haematopoietic stem cells (HSC) can differentiate into either myeloid or lymphoid lineages, which are then further differentiated into a number of different blood cell types. It is thought that the HSCs originate from an endothelial cell with haematopoietic potential (haemogenic endothelium).
Source: Reproduced and redrawn with permission from *Nature Reviews Cancer* (Speck, N.A. and Gilliland, D.G. 2002; 2: 502–13)

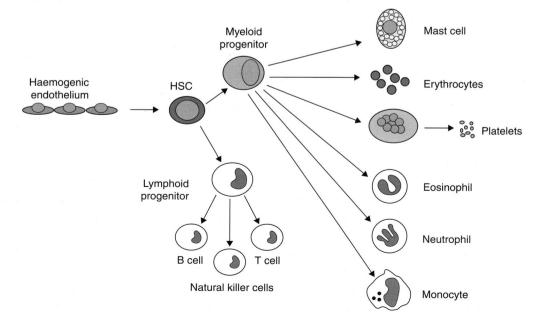

Haematopoiesis is a process that occurs in bone marrow and is responsible for the formation and development of all three blood cell types. Figure 9.1 shows the different stages of haematopoiesis. All the blood cells are derived from common *haematopoietic stem cells* (HSCs) or *haemocytoblasts*. The HSCs are also known as *pluripotent stem cells* as they can replicate themselves and have the ability to become *any* type of blood cell belonging to either the **myeloid** or **lymphoid** lineage. Myeloid stem cells can, under the influence of specific endogenous molecules, become a number of different blood cells, including RBCs, some WBCs (basophils, monocytes, and neutrophils), and platelets. Lymphoid stem cells give rise to the WBCs collectively known as lymphocytes (B cells, T cells, and natural killer cells). Thus there is a constant supply of blood cells. Read Box 9.2 to learn what happens when we donate blood.

BOX 9.2

Blood donation

Each year millions of people receive life-saving blood transfusions for a variety of diseases. The overall process of blood donation lasts approximately 1 hour, and 8–10% (approximately 1 pint or 450–500 mL) of your body's total blood volume is removed. The question you may be asking yourself is: how does the body compensate for this loss of blood? Immediately following blood donation the body starts to replenish both the plasma and cellular constituents in that 1 pint donation. The plasma levels are restored to normal within 2 days of donation. Cellular components such as RBCs require 36 days to be replaced via haematopoiesis. A person who regularly donates blood can do so every 12 to 16 weeks.

SELF CHECK 9.1

List the main cellular components of blood.

9.2 **Erythrocytes**

Erythrocytes (RBCs) are small cells that are **biconcave** in shape and contain the protein *haemoglobin* (*Hb*). RBCs are not true cells as they lack a **nucleus** and most **organelles**. They are the most numerous cells found in blood and one drop of blood may contain millions of RBCs.

Erythropoiesis is the production of RBCs, and the major steps of this process are shown in Figure 9.2. Erythropoiesis is initiated when the O_2 level within the blood decreases (hypoxia) and specialized **receptors** in the kidneys detect this. The kidneys respond to hypoxia by releasing a hormone called *erythropoietin* into the blood. Erythropoietin travels to the **red bone marrow** and stimulates it to produce more RBCs. In addition to erythropoietin, for erythrocyte production the body has to have the following nutrients: iron, vitamin B_{12}, vitamin B_6, and folic acid, as well as proteins, lipids, and carbohydrates. The main sites for erythropoiesis in adult humans are within the red bone marrow of the femur, ribs, sternum (breast bone), pelvis, and skull. In children under 5 years of age erythropoiesis occurs in all bones of the body.

The hormone erythropoietin stimulates HSCs to **differentiate** into immature RBCs know as *reticulocytes* via the series of steps shown in Figure 9.2.

1. HSCs are stimulated in the bone marrow by the hormone erythropoietin.

2. HSCs become committed to becoming RBCs by differentiating into myeloid stem cells.

FIGURE 9.2 The formation of mature erythrocytes (red blood cells) in the red bone marrow follows a series of complex events that is initiated by the hormone erythropoietin

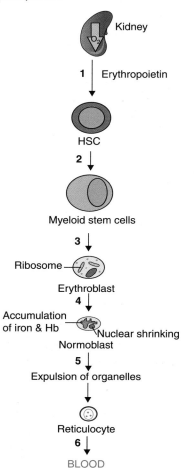

3. The myeloid stem cells, under the direction of certain factors within the bone marrow, become early *erythroblasts*, which contain ribosomes.

4. Erythroblasts start to accumulate iron and Hb. The nucleus begins to shrink in size. Erythroblasts transform into *normoblasts*.

5. Normoblasts expel organelles as more Hb accumulates within the cytoplasm.

6. Reticulocytes, immature RBCs, are released from the bone marrow into the bloodstream.

In a process that takes approximately 1–2 days, reticulocytes are slowly filled with more haemoglobin, their nucleus disappears, and they become RBCs capable of transporting O_2. In healthy adults approximately 0.8%

of circulating RBCs are *reticulocytes*. During this period of maturation reticulocytes still have RNA, required for Hb synthesis, although their nucleus has disappeared. The reticulocytes reduce in size and take on the typical biconcave disc shape of RBCs. The RBCs remain in the circulatory system for approximately 120 days before being removed by the spleen and WBCs known as macrophages. In the spleen the iron from the RBC is recycled for use in new RBCs. The amount of RBCs within the blood is dependent upon the O_2 needs of the body.

KEY POINT

Erythropoiesis is the transformation of a haematopoietic stem cell into a reticulocyte, a process that involves a reduction in cell size, an accumulation of iron, the synthesis of haemoglobin, and the loss of the cell nucleus.

SELF CHECK 9.2

Name the hormone responsible for stimulating RBC production.

Structure and function of haemoglobin

RBCs are able to carry O_2 owing to the presence of the specialized protein Hb, which is synthesized in reticulocytes in a complex series of steps. The haem part of Hb is produced in the mitochondria and cytoplasm of reticulocytes, whereas the globin part is made by ribosomes. A dietary supply of vitamin B_6, a coenzyme, is essential for Hb synthesis.

Structure of haemoglobin

Hb is made up of four polypeptide chains: two α chains and two β chains. Collectively the molecule is described as an $\alpha_2\beta_2$ tetramer. Each polypeptide chain has a haem prosthetic group with an iron atom at its centre. Each haem group can bind one O_2 molecule. Therefore there are four haem groups per Hb molecule that can bind a total of four O_2 molecules. Figure 9.3 shows the structure of the haemoglobin molecule.

FIGURE 9.3 (A) The ribbon structure and (B) the space-filling structure of the haemoglobin tetramer showing four polypeptide chains (two α chains in pink and two β chains in yellow). Each polypeptide chain surrounds a haem group (grey) containing a single iron atom
Source: Created from the Protein Data Bank, molecule 1GZX

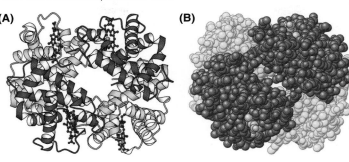

Function of haemoglobin

The primary function of Hb is to transport and distribute O_2 from the lungs to the tissues of the body. At the tissues the O_2 is exchanged for CO_2, which the blood then transports back to the lungs to be exhaled. When O_2 is bound, the haem group is red in colour (oxyhaemoglobin), and when it lacks O_2 (deoxyhaemoglobin) it is blue–red in colour.

The affinity of Hb for O_2 changes with O_2 concentration—it is high when O_2 is high (in the lungs) and low when O_2 levels are low (in the capillaries). Box 9.3 describes some of the consequences of decreased O_2 levels in blood. The affinity of Hb for O_2 is also reduced by pH—low pH causes the release of O_2 from Hb. Tissues undergoing metabolic reactions require more O_2. In these tissues there is an increase in CO_2, causing a low or acidic pH in the blood. Thus Hb is likely to release the O_2 in these tissues. In addition to O_2 Hb can transport other molecules, such as CO_2, carbon monoxide, and nitric oxide.

KEY POINT

A molecule of haemoglobin can carry four molecules of O_2.

SELF CHECK 9.3

What is the primary function of haemoglobin?

BOX 9.3

Altitude sickness

Hikers, skiers, mountain climbers, and persons who live at or near sea level who ascend to elevations above 2400 metres are at risk of developing altitude sickness or acute mountain sickness. This disorder is a result of exposure to low partial pressures of O_2 at high altitudes. If these people do not acclimatize to their surroundings they can experience a multitude of symptoms, from dizziness, nausea, and shortness of breath to the more serious symptoms of cyanosis (blue colouration of the skin), confusion, inability to breathe, and even death.

In countries such as Tibet and mountainous regions of South America the native people have adapted to living at high altitudes where O_2 levels are reduced by having a higher concentration of Hb within their blood. This allows for more O_2 to be carried in the same volume of blood as persons living at sea level, where O_2 partial pressures are much higher.

9.3 Leukocytes

Leukocytes (WBCs) are the only blood cells that are true cells as they contain both a nucleus and organelles.

WBCs make up less than 1% of the total blood volume. They are cells of the immune system that are involved in defending the body against disease and foreign material. One of the ways in which WBCs do this is via the inflammatory response (inflammation), in which they release chemicals that produce noticeable physiological effects such as redness and swelling. WBCs can be divided into five diverse groups: basophils, eosinophils, neutrophils, lymphocytes, and monocytes. All of the WBCs are produced from HSCs in the bone marrow during a process known as *leukopoiesis*, a form of haematopoiesis.

> ## KEY POINT
>
> Leukopoiesis is the term used to describe the formation of WBCs in the bone marrow.

HSCs, under the control of proteins called **cytokines**, take one of two pathways: lymphoid or myeloid. Lymphoid stem cells produce *lymphocytes* and myeloid stem cells produce *basophils, eosinophils, neutrophils,* and *monocytes*. The WBCs are further classified by structural and chemical characteristics. Eosinophils, basophils, and neutrophils are termed *granular leukocytes* or *granulocytes*, owing to the presence of cytoplasmic granules (small membrane-bound particles found in the cytoplasm). Monocytes and lymphocytes are referred to as *agranular leukocytes* or *agranulocytes* because their cytoplasm lacks granules. Mature granulocytes are stored in the bone marrow and are released into the blood when needed. Thus the bone marrow contains up to ten times more granulocytes than the blood.

Leukocytes in general do not stay in the blood for long. Granulocytes usually circulate for 4–8 hours before migration into the tissues, where they survive for 4–5 days. Monocytes can travel in the blood for 10–20 hours before migration into the tissues. Monocytes transform into macrophages in the tissues and can remain there for as long as a few years. Lymphocytes take part in long-term immunity and are able to survive from a few weeks to decades.

Basophils

Basophils are the rarest WBCs, and account for less than 1% of the total WBC population. The cytoplasm of basophils contains **histamine**, an inflammatory chemical that attracts other WBCs to sites of inflammation in the body. Basophils also produce **heparin**, an **anticoagulant** that prevents the clotting of blood.

 Histamine is covered in more detail in Section 7.2 'Histamine', within Chapter 7 of this book.

Eosinophils

Eosinophils make up approximately 2–4% of the total WBC population and are roughly the same size as neutrophils. They are easily identified by their large-lobed nucleus. Eosinophils defend the body against infection from parasites such as flatworms and roundworms by surrounding them and releasing **enzymes** that digest the parasites.

 Section 3.2 'Enzymes and enzyme inhibition', within Chapter 3 of this book, looks at enzymes in more detail.

Neutrophils

The most numerous of the WBCs are the neutrophils, accounting for 50–70% of the total WBC population. They are twice the size of RBCs. Neutrophils are also known as *polymorphonuclear neutrophils (PMN),* as they have a variety of nuclear shapes. Neutrophils are responsible for the removal of bacteria from the body, so during a bacterial infection their numbers will increase considerably. They are a type of *phagocyte*, in that they ingest the bacteria to remove it from the body.

Monocytes

Monocytes are the largest of the WBCs and make up 3–8% of the total WBC population. They are also known as macrophages. Like neutrophils, monocytes are phagocytic and are responsible for the ingestion of viruses and bacteria within the tissues of the body.

Lymphocytes

Lymphocytes make up 25% or more of the WBC population, making them the second most numerous WBC in the blood. Lymphocytes are associated with lymphoid tissues, where they play a crucial role in immunity. There are three types of lymphocyte:

1. *B lymphocytes* (*B cells*) produce **antibodies**, which bind to pathogens and target them for destruction by other immune cells such as monocytes.

2. *T lymphocytes* (*T cells*) are further subdivided into three subtypes:

   i. CD4+ (helper) T cells, which direct the immune response. These cells are important for the destruction of intracellular bacteria.

   ii. CD8+ (cytotoxic) T cells, which kill virus-infected and tumour cells.

   iii. γδ (gamma delta) T cells, which share characteristics of helper T cells, cytotoxic T cells, and natural killer cells.

3. *Natural killer* (*NK*) *cells* kill cells that have been infected by a virus or are cancerous. NK cells must receive an initiating signal in the form of an endogenous molecule before they destroy the cell.

All of the lymphocytes are produced in the bone marrow via leukopoiesis. T cells are then matured in the **thymus**.

> **SELF CHECK 9.4**
>
> Name the process that results in the formation of WBCs and list the five different types of WBC.

9.4 Plasma

Plasma is the liquid matrix of blood in which the blood cells are suspended. Plasma accounts for approximately 55% of whole blood, with a total volume of 2.7–3.0 litres in the average human adult. Plasma is a yellow-coloured aqueous solution that contains 92% water, 8% plasma proteins, and trace amounts of other dissolved materials and substances.

The primary function of plasma is to circulate dissolved nutrients such as glucose, amino acids, and fatty acids; hormones; and clotting factors. Plasma also transports waste products such as CO_2, urea, and lactic acid for excretion by the body's organs. Substances transported within plasma are bound to carrier proteins such as **albumin**, which accounts for nearly 60% of the total protein content of plasma. The components of plasma are summarized in Table 9.2.

> **KEY POINT**
>
> Plasma is the fluid matrix in which blood cells are suspended.

Plasma serum is plasma without any fibrinogen or other clotting factors present. In the laboratory, plasma is prepared by spinning a blood sample in a centrifuge to separate the blood into its component parts. Figure 9.4 shows the components of blood following centrifugation. The result is three clearly distinct layers. The upper layer is a clear yellow-coloured solution that is plasma. The middle layer,

TABLE 9.2 The components of blood plasma

Component	Examples
Albumins	Major protein constituent of plasma—total volume accounts for nearly 60% of all plasma proteins
Globulins	Hormones, lipids, and antibodies
Fibrinogen	Essential to the coagulation process, fibrinogen is converted to fibrin, which forms the stable blood clot
Other proteins	Enzymes, coenzymes, hormones etc.
Electrolytes	Sodium (Na^+), potassium (K^+), calcium (Ca^{2+}), magnesium (Mg^{2+}), chloride (Cl^-), bicarbonate (HCO_3^-), and sulphate (SO_4^-) ions are needed for cell function
Organic nutrients	Lipids, cholesterol, carbohydrates, and amino acids provide energy for cellular activities in the body
Organic waste products	Urea, uric acid, creatine, bilirubin, and ammonium ions (NH_4^+)

FIGURE 9.4 **The components of blood separated by centrifugation**

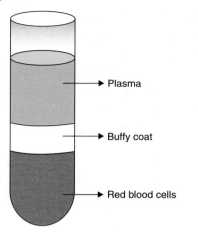

→ Plasma

→ Buffy coat

→ Red blood cells

which is white in colour, is known as the *buffy coat* and contains the leukocytes and platelets. The bottom red layer contains the RBCs.

9.5 **Blood disorders**

Blood disorders can arise from defects in blood vessels or from abnormalities in the blood itself, and include bleeding disorders, platelet disorders, haemophilia, and anaemia. Table 9.3 summarizes the normal haematological reference values for a full blood count. Values that are different from these may be indicative of disease or a blood disorder.

Blood disorders are determined by laboratory testing, using a blood sample that is usually extracted from a vein in the arm using a needle. The cellular components of the blood are evaluated using a variety of diagnostic testing to determine if an abnormality is present. In this section anaemias and leukaemias will be covered.

TABLE 9.3 **Haematological reference values for healthy adults**

Test	What the test measures	Normal range
Haemoglobin (Hb)	Amount within red blood cells (RBCs)	13.0–18.0 g/dl (m) 11.5–16.5 g/dl (f)
RBC	Number of RBCs in a specific volume of blood	4.3–5.7 × 10^{12}/l (m) 3.9–5.0 × 10^{12}/l (f)
Mean corpuscular volume (MCV)	Average size of an individual RBC	80–103 fl/cell
Mean corpuscular haemoglobin (MCH)	Average amount of oxygen-carrying Hb in a RBC	27–32 pg/cell
Haemocrit (HCT)	Proportion of total blood volume made up of RBCs	39–50% (m) 36–45% (f)
Platelets	Number of platelets in a specific volume of blood	1.5–4.5 × 10^{11}/l
WBC	Number of WBCs in a specific volume of blood	4.0–11.0 × 10^{9}/l
Plasma viscosity (25°C)	Measurement of the water and nutrient content of blood plasma	1.50–1.72 mPa · s
Reticulocytes	Number of reticulocytes in a specific volume of blood	30–125 × 10^{9}/l

dl, decilitre; f, female; fl, **femtolitre**; g, gram; l, litre; m, male; mPa · s, millipascal seconds; pg, picogram.

Anaemia

Anaemias are often characterized by a deficiency in RBCs. The definition for anaemia is reduced O_2-carrying capacity of blood and thus anaemia is a symptom and not a disease per se. Reduced iron is the most common cause of anaemia in the UK, and is referred to as iron-deficiency anaemia. This reduction in iron directly affects the production of haem molecules, thus reducing the O_2-carrying capacity of RBCs.

Anaemias can either be acquired or inherited, and anyone can have anaemia. It is most common in women of childbearing age, men and women aged over 75 years, and children and adolescents. Usually this is because their diets are deficient in nutrients such as iron, folic acid, and vitamin B_{12}. As people age they become less able to absorb vitamin B_{12} directly from the food they eat and become anaemic. This may be the result of an underlying autoimmune disease that damages the parietal cells in the stomach, known as *pernicious anaemia*. A rare form of anaemia known as *aplastic anaemia* is caused when there is damage to the bone marrow and new RBCs cannot be produced.

People with chronic diseases such as cancer, kidney failure, and rheumatoid arthritis can also develop anaemia. Cancer cells that invade the bone marrow and replicate out of control induce anaemia by inhibiting normal erythropoiesis. Blood loss due to tumour size and cancers of the liver and kidneys may reduce the availability of erythropoietin for new RBC formation. Anticancer treatments, such as chemotherapy, are designed to stop the multiplication of cancer cells in the body. One drawback of these treatments is that they are non-specific, and often also destroy blood cells within the bone marrow.

> **KEY POINT**
>
> The common symptoms of anaemia are lethargy, weakness, dizzy spells, and feeling faint.

A person can go many months before experiencing any of the symptoms associated with anaemia. Some medicines are thought to be linked to anaemia, as shown in Case Study 9.1. If the anaemia is

---

Naina suffers with rheumatoid arthritis (RA) and has been taking ibuprofen for a few months. She went to the GP as she was feeling very tired, a bit moody, and was having some problems with her bowels. The GP arranged for her to have a blood test done and the results showed that she had a subnormal count of red blood cells (RBCs).

REFLECTION QUESTIONS

1. What is the most likely diagnosis and treatment of Naina's condition?

2. What are NSAIDs and why are they used in the management of RA?

3. Are the symptoms Naina is experiencing a direct result of taking NSAIDs for her rheumatoid arthritis?

Answers

1 The low red blood cell count means that there are less than normal levels of RBCs in the circulation. This, combined with the feeling of fatigue Naina was experiencing, suggests the most likely diagnosis is that she is anaemic. Treatment options for anaemia are dependent upon the cause. Patients may be prescribed iron supplements or given drugs such as epoetins for more chronic cases.

2 NSAIDs are non-steroidal anti-inflammatory drugs. The most common examples are aspirin and ibuprofen. They are used for treating inflammation, moderate pain, and fever; symptoms that are associated with disorders such as rheumatoid arthritis. They are usually prescribed for symptomatic relief for conditions such as RA where pain and inflammation are present.

3 The symptoms experienced by Naina, bowel problems and subnormal RBC count, are a side effect of NSAIDs. NSAIDs can affect the GI tract, and when taken on an empty stomach can result in bleeding leading to low RBC count, and an irritable bowel.

not diagnosed and allowed to progress, the severity of the symptoms increases leading to palpitations, headaches, sore gums, and shortness of breath. Treatment of anaemia is dependent upon the cause. Doctors may prescribe iron supplements or recommend iron-rich foods to reverse anaemia caused by lack of dietary iron. In the case of anaemia caused by chronic diseases such as cancer, patients are given drugs known as *epoetins* (recombinant human erythropoietin). These drugs, administered intravenously, stimulate erythropoiesis in the bone marrow, leading to increased numbers of RBCs and improved O_2-carrying capacity of the blood.

> ### SELF CHECK 9.6
>
> What is anaemia?

Sickle cell anaemia

Sickle cell anaemia (SCA) is a hereditary blood disease that results from a single amino acid mutation in the Hb gene, meaning that people with sickle cell anaemia have an abnormal type of Hb (HbS) present in their RBCs. HbS can cause the RBC to lose its characteristic biconcave shape, becoming crescent- or sickle-shaped, as shown in Figure 9.5. This change in shape affects the ability of the RBC to pass through small blood vessels such as capillaries. The result is less O_2 being transported to the tissues that require it.

FIGURE 9.5 A schematic representation of a normal and sickled red blood cell (RBC). RBCs that are affected by sickle cell anaemia are crescent shaped when compared with normal biconcave RBCs

Normal red blood cell

Sickled red blood cell

> ### KEY POINT
>
> Sickle cell anaemia results in less O_2 being transported to the tissues that require it.

 Hereditary disorders are described in more detail in Section 2.5 'What can happen when things go wrong?', within Chapter 2 of this book.

In the UK approximately 300 babies are born each year with SCA. SCA is diagnosed via a full blood count, which reveals haemoglobin levels in the normal range but a high reticulocyte count. The body tries to compensate for the sickle cells by producing more RBCs, which may lead to the development of *haemolytic anaemia*.

SCA is managed from birth to death via a daily dose of folic acid. For the first 5 years of life **penicillin** or similar antibiotics are given daily owing to the immaturity of the immune system, which makes the children more prone to developing childhood diseases. Another form of treatment would be a bone marrow transplant.

Leukaemia

Leukaemia is a **malignant** condition which affects the formation of WBCs in the bone marrow. It is a broad term used to describe a range of diseases that prevent WBCs from undergoing normal leukopoiesis in the bone marrow. Instead, immature WBCs remain in the bone marrow and reproduce in an uncontrolled manner. These abnormal WBCs take over the bone marrow and prevent haematopoiesis. This leads to a shortage in the number of normal blood cells such as RBCs, WBCs, and platelets. Haematopoiesis and leukopoiesis have been covered in Sections 9.1 and 9.3 of this chapter, respectively.

> ### KEY POINT
>
> Leukaemias are malignant diseases that prevent the formation of mature differentiated WBCs.

In the UK, approximately 7000 people are diagnosed with leukaemia each year. It affects both adults and children and is the most common childhood cancer.

Leukaemias are divided into two main forms: *acute* and *chronic*. These forms are further subdivided according to the type of WBC that is affected: *lymphoblastic (lymphocytic) leukaemias* and *myeloid (myelogenous) leukaemia*. The combination of these two classifications gives four main categories of leukaemia.

Acute leukaemia

Acute leukaemia is a rare form of cancer; however, its progress in the body is rapid, involving immature WBCs. There are two main types of acute leukaemia: *acute lymphoblastic leukaemia (ALL)* and *acute myeloid leukaemia (AML)*. ALL affects the WBCs known as lymphocytes, whereas AML affects the myeloid cells of the body. In the UK an estimated 2400 new cases of acute leukaemia are diagnosed each year. Of these, 1800 are AML. AML is more common in older people, with the majority of cases occurring in persons 50 years and older.

> **KEY POINT**
>
> Acute leukaemia is classified according to the type of white blood cell that is affected.

The incidence of acute leukaemia is higher in males than females. The reason for this is not fully understood. The causes of this type of leukaemia are not well known, but risk factors linked to this disease are exposure to:

- Radiation, which occurs in two forms: natural and man-made. For this type of illness it is man-made radiation that poses the greatest risk, examples of which are: X-rays, old-fashioned cathode-ray televisions, nuclear medicine, and building materials.

- Benzene, which is found in cigarettes and is a chemical used in the manufacturing industry for products such as paints, plastics, dyes, and pesticides.

Diagnosis of acute leukaemia is by a blood test. It is common for the acute leukaemia cells to be seen within blood smears. Additional testing is done by a bone marrow test to identify the affected WBC.

The current treatment options available for acute leukaemia are a combination of **chemotherapy** and

BOX 9.4

Chemotherapy

Chemotherapy is the term used to describe anti-cancer drugs that destroy cancerous cells within the body by inducing programmed cell death (apoptosis). Chemotherapy drugs can be divided into cytotoxic chemotherapy and targeted treatments. Cytotoxic chemotherapy is non-specific and destroys both normal and cancerous cells. Targeted treatments are newer drugs that have been designed to specifically target a certain part of the cancer cell.

To date, 200 different types of cancer have been described and there are 50 available cytotoxic chemotherapy drugs used in treatment regimes. The drugs can be administered individually or collectively as combination therapy. The type of treatment is dependent on the type of cancer, the location of the cancer in the body, and whether the cancer has metastasized (spread) to another region in the body.

Examples of chemotherapy drugs currently used in the UK include cisplatin (testicular, bladder, lung, oesophagus, stomach, and ovarian cancers); etoposide (lung, ovarian, and testicular cancers); fluorouracil or 5FU (bowel, breast, skin, stomach, and oesophagus); oxaliplatin (bowel and oesophageal cancers); and temozolomide (brain cancer). Cytotoxic chemotherapy is usually administered intravenously or orally, although other routes of administration are possible, such as topical creams for skin cancer.

radiotherapy. Some of the drugs used in chemotherapy are mentioned in Box 9.4. In certain cases a bone marrow transplant may be performed for those who do not respond to chemotherapy or radiotherapy. The overall outlook for people with acute leukaemia is varied owing to the many subtypes of the disease. Some of the subtypes are more challenging to treat than others. Young people often have a better chance of recovery than older people.

Chronic leukaemia

Chronic leukaemia develops over many years and does not require immediate treatment. It is classified according to the type of WBC that is affected, there being two main types: *chronic lymphocytic leukaemia (CLL)* and *chronic myeloid leukaemia (CML)*. CLL is cancer of the lymphocytes and CML is cancer of the myeloid cells.

CLL occurs mainly in older people over the age of 65 years; it can develop in children but this is a rare occurrence. Statistics show that CLL mainly affects males. In the UK there are approximately 2100 new cases of CLL each year. The overall outlook for CLL patients is poor, with approximately 50% of cases living for only 5 years following diagnosis.

CML is a very rare form of leukaemia, with approximately 600 new cases being diagnosed each year in the UK, and is more common in people aged 40–60 years. CML is curable via bone marrow transplantation but this treatment option is not suitable for all people with the disease. The outlook for people with CML is hopeful owing to the introduction in 2001 of *imatinib*, the first in a class of inhibitors of the abnormal tyrosine kinase enzyme believed to be involved in the development of CML. Early research shows that people can live for up to 20 years after a diagnosis of CML.

 Enzyme inhibitors are covered in more detail in Chapter 3.

SELF CHECK 9.7

Define the term leukaemia and name the different types of acute and chronic leukaemia.

9.6 Haemostasis

Haemostasis, the process that prevents excessive loss of blood, is essential for maintaining life. The loss of blood or bleeding can be detrimental to life and needs to be kept under control. Haemostasis occurs in four distinct phases:

1. *Vascular phase*—injury/damage to a blood vessel causes constriction of the blood vessel. This constriction slows down blood flow within the injured vessel.

2. *Platelet phase*—injury to the endothelial lining of the blood vessel activates platelets. Platelets stick to the endothelium and recruit more platelets to the area to form a '*platelet plug.*'

3. *Coagulation phase*—the platelet plug is converted to a stable blood clot known as a *fibrin clot*. The formation of a fibrin clot is the result of a complex series of events. These events are regulated by a series of plasma constituents known as *coagulation factors*.

4. Fibrinolysis—the blood clot is broken down and removed when tissue repair of the damaged vessel is complete.

Formation of platelet plugs

Platelets, also known as *thrombocytes*, are non-nucleated disc-shaped cells that circulate freely in the blood. They are formed in the bone marrow by a process known as *thrombopoiesis* from *megakaryocytes,* as shown in

FIGURE 9.6 The process of platelet production in the bone marrow—thrombopoiesis.
Source: Reproduced and redrawn with permission from *Blood* (Drachman, J.G. Inherited thrombocytopenia: when a low platelet count does not mean ITP. 2004; 103: 390–8)

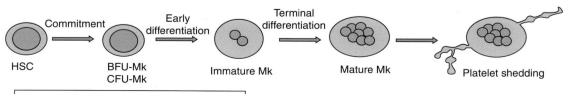

9 Haematology

180

figure 9.6. Megakaryocytes are produced by differentiation from HSCs during haematopoiesis (see Figure 9.1 for the process of haematopoiesis). Once the megakaryocyte has matured it is able to produce platelets.

Thrombopoietin, a hormone produced in the liver and kidneys, regulates the production of platelets in the bone marrow. Thrombopoietin stimulates the mature megakaryocyte to form new platelets by budding off from the megakaryocytes, as shown in Figure 9.6. Platelets are the smallest and second most numerous blood cells and are present in the blood at 1.5–4.5×10^{11} cells/litre. The average lifespan of a platelet is 8–12 days and each megakaryocyte can produce between 5000 and 10 000 platelets per day. Platelets are removed and destroyed in the spleen and liver.

Platelet activation occurs in two steps, activation and aggregation. This process requires a series of both intra- and extracellular biochemical signalling pathways. For platelets to become active they need to make contact with the injured endothelium of the blood vessel. Contact is made via receptors on the surface of the platelet attaching to substances such as **von Willebrand factor** within the damaged endothelium. The platelets will then stick to the damaged blood vessel injury site. The initial step of a platelet adhering to the endothelium triggers the biochemical signalling events that change the shape of the platelet. The platelet goes from discoid to 'activated star' shape. Once activated, platelets release cytokines that attract more platelets to the injury site by a process known as **chemotaxis**. The attracted platelets spread out over the entire surface of the injury site to form a 'platelet plug'. The platelet plug seals the damaged vessel to prevent any further blood loss. Figure 9.7 shows the events of platelet activation and aggregation.

> **KEY POINT**
>
> Platelets are central to the process of clot formation in response to injury.

Antiplatelet therapies

In certain disease states, such as cardiovascular disease, platelet plugs are formed in the arterial blood vessel wall when circulating platelets adhere to glycated

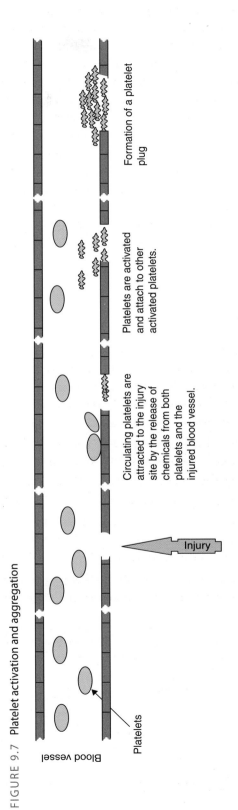

FIGURE 9.7 Platelet activation and aggregation

Formation of a platelet plug

Platelets are activated and attach to other activated platelets.

Circulating platelets are attracted to the injury site by the release of chemicals from both platelets and the injured blood vessel.

Injury

Platelets

Blood vessel

endothelial cells and in turn recruit more platelets. This can be serious and can lead to death if the blood is not able to flow freely through the blood vessels. Therapeutic intervention is required to redress this imbalance and restore normal haemostasis. Antiplatelet therapies are divided into three groups: aspirin, thienopyridines, and glycoprotein IIb/IIIa inhibitors.

Aspirin

Aspirin, a member of the non-steroidal anti-inflammatory family of drugs, has antiplatelet effects. It inhibits the formation of a platelet plug by interfering with the actions of the enzyme cyclooxygenase. Aspirin is usually prescribed as a secondary prevention measure following a cardiovascular event, in particular **coronary artery disease**.

Thienopyridines

Thienopyridines are **adenosine-5′-diphosphate (ADP)** receptor **antagonists**. They exert their antiplatelet effects by preventing ADP from binding to its receptors on the platelet cell surface. They are slightly more effective in inhibiting platelet aggregation than aspirin, and are usually administered to patients with advanced **atherosclerosis**, who are at risk of developing a heart attack. Thienopyridines are also given after coronary angioplasty and/or coronary stent implantation. The thienopyridines that are used in the UK are clopidogrel, ticlopidine, and prasugrel.

 Section 4.2 'Basic concepts of pharmacodynamics', within Chapter 4 of this book, outlines different receptors in more detail.

Glycoprotein IIb/IIIa inhibitors

Glycoprotein IIb/IIIa is a receptor found on the platelet cell surface that is triggered by the action of molecules such as ADP. It is responsible for platelets adhering during platelet plug formation. Thus inhibitors of this receptor prevent platelets from attaching to each other. At present there are three *glycoprotein IIb/IIIa inhibitors*: abciximab (a **monoclonal antibody**), eptifibatide, and tirofiban (both non-peptide antagonists).

In addition to these drugs there is a novel phosphodiesterase inhibitor, cilostazol, that has both antiplatelet and **vasodilating** properties.

Coagulation: blood clotting

Coagulation is the complex process by which blood forms clots. Coagulation plays an important role in haemostasis by halting blood loss from an injured blood vessel. Coagulation begins immediately following injury to the endothelium of a blood vessel.

Blood clotting requires the use of plasma proteins called *coagulation factors*. These are involved in a complex series of events that form fibrin strands which stabilize the platelet plug. The major coagulation factors are shown in Table 9.4. Coagulation is classically divided into three pathways: the *intrinsic pathway,* the *extrinsic pathway,* and the *final common pathway,* as shown in Figure 9.8.

Approximately 750 000 people in the UK take warfarin, an anticoagulant often referred to as a 'blood thinner'. This oral medication prevents the process of coagulation, and is given to people who are at risk of developing a blood clot or need treatment for an existing blood clot. Warfarin exerts its anticoagulant effects by antagonizing the actions of vitamin K in the coagulation cascade. It is prescribed for the treatment and prevention of the following conditions:

- Deep vein thrombosis—blood clots in the veins of the legs.
- Pulmonary embolism—blood clots in the lungs.
- Ischaemic stroke—blood clot in the carotid artery that supplies the brain with blood.
- Heart attack or myocardial infarction—blood clots in the coronary vessels of the heart.

> **KEY POINT**
>
> Anticoagulant therapies are different to antiplatelet therapies, in that anticoagulants target blood clots that have formed in veins, as opposed to clots 'plugs' in the arteries that are primarily composed of platelets. It should be noted that, clinically, anticoagulants carry an increased risk of bleeding compared with antiplatelet therapies.

Haemophilia

Haemophilia is a condition in which essential clotting factors (see Table 9.4) are absent from the blood.

FIGURE 9.8 The coagulation cascade

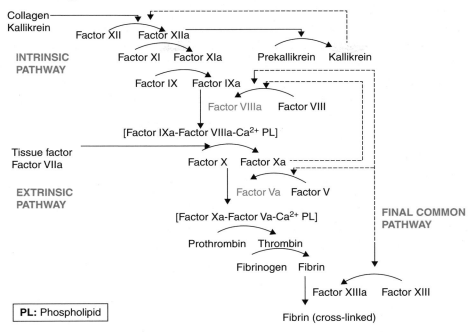

PL: Phospholipid

There are two types of haemophilia, type A (lack of factor VIII), which represents 90% of cases, and type B (lack of factor IX). People with haemophilia are unable to form blood clots and this can lead to serious health problems if not controlled. *Factor concentrate therapy* is the primary treatment option for both type A and B haemophiliacs. The disease is classified into three categories: mild, moderate, and severe. Treatment options are dependent on the category of the disease. People with mild haemophilia can be prescribed desmopressin, which stimulates the release of factor VIII. It is also possible to take antifibrinolytics such as tranexamic acid and aminocaproic acid to prevent the breakdown of blood clots. See Integration Box 9.1 for further information on haemophilia.

TABLE 9.4 Some of the factors involved in the coagulation cascade

Factor/substance	Function
I (fibrinogen)	Converted to fibrin to form the blood clot
II (prothrombin)	Active form (IIa) activates other coagulation factors, protein C, and platelets
VII	Enzyme that causes blood clot formation
VIII (antihaemophilic factor)	Cofactor of factor IX, with which it forms the tenase complex
IX (Christmas factor)	Activates factor X after it has formed the tenase complex with factor VIII
XI	Enzyme of the coagulation cascade
XII (Hageman factor)	Activates factor XI and prekallikrein
von Willebrand factor	Binds to factor VIII and mediates platelet activation

Fibrinolysis

Fibrinolysis is the term used to describe the breakdown of the fibrin clot formed during coagulation. The enzyme **plasmin** aids in the breakdown of the blood clot by cutting the fibrin network at various places. Plasmin is produced in the liver in its inactive form, *plasminogen*, which is subsequently activated by the enzymes urokinase and tissue plasminogen activator (tPA). The blood clot is turned into smaller pieces and removed from the blood by other enzymes known as **proteases**. Blood clot fragments can also be removed from the blood by the liver and kidneys, both of which act like sieves to filter the blood clot pieces from the circulating blood. Fibrinolysis can be divided into two types: primary fibrinolysis is the normal breakdown of blood clots in the body whereas secondary fibrinolysis is the breakdown of blood clots owing to a blood disorder, medicines, or other causes. The process of fibrinolysis involves a series of complex steps that are outlined in Figure 9.9.

Drugs that initiate fibrinolysis (thrombolytics) are given to people following events such as a heart attack. These drugs dissolve any blood clots within the coronary artery that may have caused the heart attack. Examples of thrombolytic agents are: streptokinase, urokinase, and alteplase.

FIGURE 9.9 Fibrinolysis: the process of blood clot breakdown

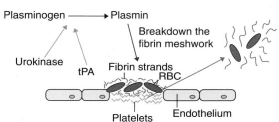

SELF CHECK 9.8

Name the four phases of haemostasis and explain why platelets are important for the maintenance of haemostasis.

9.7 **Blood groups**

There are more than twenty genetically determined blood group systems, with the *ABO* and *rhesus (Rh)* systems being the most important ones used for blood typing for transfusion. Not all blood groups are compatible with each other, and mixing of incompatible blood groups can lead to **agglutination** or clumping, which can result in **haemolysis** or even death.

ABO blood group system

There are four principal blood group **phenotypes** present within the human population: *A*, *B*, *AB*, and *O*. The differences between the individual blood groups is due to the presence or absence of specific proteins known as **antigens**, found on the RBC cell membrane, and antibodies within the blood plasma. Individuals within a population will have different types and combinations of these antigens and antibodies and we inherit these from our parents. The ABO gene is **autosomal** (not on either of the sex chromosomes), located on chromosome 9. A and B blood groups are dominant over type O blood group, with A and B being **codominant**. Each person has two copies of the genes coding for their ABO blood group.

 Section 2.5 'What can happen when things go wrong', within Chapter 2, discusses autosomal inheritance in more detail.

Based on the ABO blood group system there are four blood types:

- *Blood group A*—Type A antigens are present on the surface of RBCs and B antibodies are in blood plasma. Type A individuals can receive blood from other type A and type O individuals.

- *Blood group B*—Type B antigens are present on the surface of RBCs and A antibodies are in blood plasma. Type B individuals can receive blood from other type B and type O individuals.

- *Blood group AB*—Types A and B antigens are present on the surface of RBCs and neither A nor B antibodies are in blood plasma. Type AB individuals are *universal receivers* as they can receive blood from all other blood groups because they do not produce antibodies to either type A or type B antigens.

- *Blood group O*—Neither type A nor type B antigens are present on the surface of RBCs, but both A and B antibodies are present in blood plasma.

People with a type O blood group can donate to any of the other blood groups (*universal donors*) without rejection, owing to the absence of type A and type B antigens. However, they can only receive blood from other type O individuals.

The rhesus (Rh) system

The Rh system or Rh factor is named for the Rh antigen, designated by the letter D, which was first identified in rhesus monkeys (hence the name). The Rh antigen is present or not present on the surface of RBCs and those individuals who have Rh antigens are referred to as Rh positive (Rh+) while those lacking the antigen are Rh negative (Rh−). Approximately 85% of the population are positive for the Rh antigen.

For the purpose of ensuring compatibility when a person receives a blood transfusion the two typing systems are combined. For example, people who have blood group A and have the Rh antigen are classified as A+ (A positive). Those who do not have the Rh antigen are A− (A negative). This applies for the remaining blood groups: B, AB, and O. Rh+ individuals can only donate blood to individuals who also have the Rh factor and whose ABO blood groups are also compatible, while individuals that lack the Rh factor can donate blood to both Rh+ and Rh− individuals.

> **KEY POINT**
>
> According to the two blood grouping systems (ABO and Rh), a person can belong to any one of the eight blood groups: A +, B +, AB +, O +, A −, B −, AB −, or O−.

> **SELF CHECK 9.9**
>
> Taking the two blood grouping systems (ABO and Rh) into account, draw up a table of blood transfusion compatibility for all eight blood groups.

CHAPTER SUMMARY

➤ Blood is composed of four main components: red blood cells (RBCs), white blood cells (WBCs), platelets, and plasma. Haematopoiesis is the formation of blood cells in the bone marrow.

➤ RBCs are formed via a process known as erythropoiesis in the red bone marrow. The primary function of RBCs is to transport oxygen from the lungs to the tissues and cells of the body.

➤ Haemoglobin is a protein found within RBCs that allows for oxygen transportation within the blood. Anaemia is a blood disorder that affects the oxygen-carrying capability of RBCs.

➤ Iron-deficiency anaemia is the most common cause of anaemia in the UK.

➤ Leukopoiesis is the formation of WBCs. There are five types of WBC: basophils, eosinophils, lymphocytes, monocytes, and neutrophils.

➤ Plasma is the liquid matrix in which blood cells are suspended. Albumin is the major protein found in plasma.

➤ Platelets, the smallest of the blood cells, are formed during thrombopoiesis. They are important to the body because they regulate haemostasis in the blood. Platelets, along with coagulation factors, prevent blood loss from vessels by the formation of clots.

➤ Non-steroidal anti-inflammatory drugs such as aspirin are antiplatelet drugs that inhibit the cyclooxygenase enzyme and thus the formation of the platelet plug.

➤ Antiplatelet therapies target platelet plugs in the arteries whereas anticoagulants dissolve clots in veins.

➤ Fibrinolysis is the process whereby blood clots are broken down to prevent restriction of blood flow in the vessels.

➤ Haemophilia is an inherited condition in which patients cannot form blood clots owing to the absence of clotting factors.

➤ Blood groups are determined by the presence or absence of specific protein molecules found on the plasma membrane of RBCs. There are eight possible blood groups based on combinations of both the ABO and Rh systems.

FURTHER READING

Graf, T. Immunology: blood lines redrawn. *Nature* 2008; 452: 702–3.

This article provides an in-depth description of the cells found within blood.

Foley, R.N. Emerging erythropoiesis-stimulating agents. *Nature Reviews Nephrology* 2010; 6: 218–23.

An interesting review on the subject of erythropoiesis and pharmacological agents that stimulate the process.

Burnett, A., Wetzler, M., and Löwenberg, B. Therapeutic advances in acute myeloid leukaemia. *Journal of Clinical Oncology* 2011; 29: 487–94.

An in-depth insight into the emerging treatment options for acute myeloid leukaemia.

NHS Choices website. <www.nhs.uk >

A good resource for exploring the science of haematology.

Bain, B.J. *Blood Cells: A Practical Guide*. 4th edn. Blackwell, 2006.

This text provides a comprehensive review of the blood cells.

This is just the beginning

ELSIE E. GASKELL

Pharmacy involves the study of therapeutic molecules and their clinical uses. This book has introduced you to some aspects of the biology of cells and the physiology of the human body; it has outlined the relevant concepts of biochemistry, and addressed basic principles in pharmacology. It has aimed to give you an insight to the multifaceted science that is therapeutics. Now you need to build on this knowledge to fully understand how the body works in health and disease and therefore appreciate how drug molecules can have curative effects. To achieve this, however, you cannot solely rely on therapeutics. Only with continuous reference to its associated chemical-, physical-, and practice-oriented disciplines will you truly understand the essence of pharmacy as an interdisciplinary profession.

Learning objectives

Having read this chapter you are expected to be able to:

➤ Appreciate the required integrated approach to fully understand disease states and therapeutic interventions.

10.1 Where do you go from here?

This book provides an introduction to the therapeutics topics you will encounter in more depth and breadth in future years on your pharmacy and related programmes. In the first year of your programme the topics covered in this book may be introduced under varying titles, including cell biology, biochemistry, introduction to physiology and pharmacology, and introduction to the scientific basis of therapeutics, among others, but they all aim to provide a sound foundation upon which you can build your future learning. There are topics that you will encounter in your future studies that are not considered here, for example immunology and infection, or have only briefly been introduced, for example the cardiovascular system. Nonetheless, the layout and style of the chapters should enable you to link them and recognize the complementary nature of pharmaceutical sciences and clinical practice.

Regardless of the apparent structure of your particular programme of study, a good starting point for your studies is always to look for the interrelated connections. Our bodies are complex and integrated systems, and pharmacy is concerned not only with the biology but is reliant on an understanding of the chemical and physical properties of the therapeutic agent as well as the practice elements of the profession. We will consider next an integrated view of a disease state. You can do this for any specific topic during your studies.

> **KEY POINT**
>
> For you to understand the relevant processes and take advantage of this knowledge, an integrated approach to your learning is essential.

10.2 An integrated approach to disease states

The body constantly aims to maintain a continuous equilibrium between its internal and external environment by regulating numerous systems and processes. Disease states often occur as a consequence of these equilibria being disrupted, or indeed the homeostasis of a system may be disrupted as a direct or indirect consequence of a disease state, explaining the associated symptoms. Let's consider the common disease diabetes.

Diabetes mellitus is a group of metabolic disorders characterized by high levels of sugar in the blood. There are various causes for the consequential health problems associated with diabetes, which explains the different treatment options available. To fully comprehend the clinical aspects of diabetes and consider the treatments available we first have to fully appreciate the physiological processes that occur in the body.

As described in Chapter 3, under normal circumstances our body harvests the energy stored in the food and drink we consume to facilitate the everyday functioning of its cellular and subcellular processes. A high-energy compound in the centre of the metabolic processes within the body is *glucose*, a monosaccharide (simple sugar). When required by the body, glucose is metabolized to yield molecules of adenosine-5′-triphosphate (ATP) via glycolysis and the citric acid cycle (also referred to as the Krebs cycle). The body uses the ATP as energy to support other vital processes. The glucose in the body can also be stored as glycogen in the liver when supplies are in excess,

via the process of glycogenesis. A schematic diagram summarizing the metabolism of glucose is presented in Figure 10.1. Here you can see how different food sources: complex carbohydrates, proteins, and fats are fed into the metabolic pathways.

 Section 3.3 'Metabolic pathways and abnormal metabolism', within Chapter 3, covers the metabolism of carbohydrates, proteins, and fats in detail.

Glucose is a vital molecule required by our bodies to maintain a healthy state. Our bodies strive to balance the different interlinked metabolic processes to maintain a set amount of glucose circulating in the body. We have learnt in Chapter 8 that when the blood sugar level drops hypoglycaemia develops, and when there is a prolonged increased level of glucose in the blood (hyperglycaemia) serious health issues arise. Diabetes develops when the homeostatic control of circulating glucose level is compromised and levels continuously exceed those tolerated by the body.

 See Section 8.3 'Major homeostatic mechanisms in the human body', within Chapter 8, for more information on blood glucose homeostasis.

After a meal our body starts digesting the food. This is an involuntary process facilitated by the autonomic nervous system (ANS). Not only is the ANS involved in moving the food into the stomach and through the intestines, it also stimulates the mechanical and chemical breakdown of the food, allowing absorption of the nutrients from the food into the bloodstream.

FIGURE 10.1 Schematic diagram of glucose metabolism

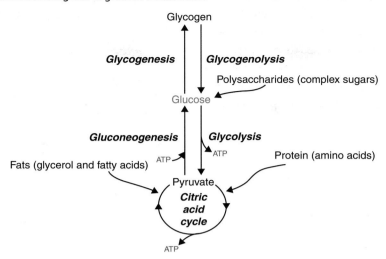

The blood carries the absorbed nutrients from the gastrointestinal tract to the liver to be filtered and further processed before distribution to other organs in the body. Blood contains of a number of different components and has varying roles. For the distribution of nutrients, however, it is the blood plasma that enables the delivery of the absorbed sugars, amino acids, and fatty acids to cells for nourishment. Of all the nutrients carried by the blood plasma, the level of glucose (measured as blood sugar level) is the most regulated.

 Blood, its composition and function are detailed in Chapter 9.

A range of organs are involved in the complex process of regulating the freely circulating glucose level but the pancreas and the liver have central roles. The pancreas is an **endocrine gland** with a central role in glucose regulation. It monitors the level of glucose in the blood and secretes different hormones in response to the detected levels:

- *Insulin* is secreted by the β-cells in the pancreatic islets of Langerhans in response to high levels of glucose in the blood and stimulates its uptake by the target cells and stores it as glycogen.

- *Glucagon* is secreted by the α-cells in the pancreatic islets of Langerhans in response to decreased levels of glucose in the blood and stimulates the output of glucose from the target cells into the bloodstream.

 The pancreas and hormonal regulation of glucose are introduced in Section 7.6 'Hormones', within Chapter 7.

It is the fine interplay between glucagon and insulin that maintains blood glucose homeostasis. These hormones, along with others which have recognized roles in regulating glucose levels, are secreted into the bloodstream and travel, together with the dissolved nutrients, to the recipient cells throughout the body. Cells have membranes that surround them and house the vast number of proteins acting as receptors for specific endogenous molecules. The type and quantity of the receptors expressed are dependent on the cell type. The membrane therefore regulates the trafficking of nutrients and ions into the cell. The hormones involved in the regulation of circulating sugar levels interact with their specialized receptors, an action which triggers a sequence of events inside the cell; these hormones do not cross the membrane and do not enter the cell cytoplasm themselves.

The biology of cells is introduced in Section 2.1 'The unit of life: the cell', within Chapter 2.

The glucose uptake system of the target cell is principally driven by a receptor-mediated system. The insulin acts as a signal, binding to the insulin receptors (a tyrosine kinase receptor) found in the membranes of target cells in skeletal muscles and **adipose tissue**. This triggers a sequence of events that results in **vesicles** containing specific intracellular molecules (the glucose transporters) migrating to the cell membrane surface where they can take up the glucose surrounding the cells. The outcome is that the glucose is removed from the blood plasma and either further metabolized by the muscle cells to yield energy or converted to glycerol and used to form triglycerides in adipocytes.

Glucagon, on the other hand, results in the opposite outcome. It is present in the bloodstream when circulating levels of glucose are low and binds to its specific glucagon receptors (a G-protein-coupled receptor) in the cell membranes of hepatocytes, kidney cells, and adipocytes. This binding triggers a sequence of events leading to changes in the concentrations of a number of second messengers and **phosphorylation** of key enzymes in the glucose metabolism pathways, resulting in the stimulation of glycogenolysis and gluconeogenesis as well as inhibition of glycogenesis and glycolysis. This results in an increased output of glucose from these cells into the bloodstream.

 Section 4.2 'Basic concepts of pharmacodynamics', within Chapter 4 details the different types of receptors involved in signal recognition.

The liver has a range of functions in the body, one of which is involvement in the regulation of glucose levels in the blood. The ANS, via the liver, affects the level of free glucose by regulating the processes of gluconeogenesis, glycogenesis, and glycogenolysis (refer to figure 10.1). This is achieved by influencing key enzymes involved in these metabolic pathways. The two branches of the ANS, the parasympathetic and sympathetic nervous systems, are proposed to have opposing functions on the hepatic metabolism of glucose. The parasympathetic (often simplified to the 'rest and digest' system) promotes glucose uptake by the liver and its storage as glycogen (glycogenesis). Conversely, the sympathetic nervous system (often simplified to the 'flight or fight' system) promotes the breakdown of glycogen (glycogenolysis) and the synthesis of glucose (gluconeogenesis). This results in the output of glucose from the liver into the bloodstream, making it available for further metabolism.

 The autonomic nervous system is covered in more detail in Chapter 6.

Other organs central to the maintenance of blood glucose levels are the adrenal gland and the adipose tissue. The overall regulation of glucose levels in the blood is a very complex process and continuous research on the topic is further elucidating the involvement of the various body systems. The overarching control is, however, from the central nervous system, via the brain and the afferent and efferent nerves of the peripheral nervous system. The responses from the higher centres of the central nervous system are presumed to affect a number of organs, including the pancreas, adrenal gland, skeletal muscles, **adipocytes,** and the liver.

 An introduction to the nervous system is presented in Chapter 5.

The hormonal and neural control of blood sugar levels results in tight regulation of the blood sugar level. The complexity of the whole system is considerably deeper than summarized here, and indeed we may not yet fully understand the specific details of all the processes involved. As indicated, a disease state often develops when the normal processes within the body are disrupted. For diabetes, a number of malfunctions may be the cause of the clinical symptom of a chronically elevated glucose blood level. Because of this, diabetes is typically classified into two main types: type 1 and type 2, although there are other types such as **gestational diabetes**.

- Type 1 diabetes results from an **autoimmune disorder** in which the insulin-producing β-cells in the islets of Langerhans are destroyed and consequently the individual does not have sufficient hormonal control over the level of glucose.

- Type 2 diabetes is caused by the development of impaired tolerance of the insulin receptors to the circulating hormone, or the reduced production of insulin. In either case, the regulation of blood sugar becomes inadequate.

Treatments available for people diagnosed with diabetes are varied and the choice of medication depends on the type and stage of the diabetes as well as preferences for the treatment regimens. Type 1 diabetic patients and some type 2 patients require regular insulin injections to help regulate their blood sugar level. For some type 2 diabetic patients, medication that lowers the blood glucose level is prescribed in addition to advice on a healthy diet and active lifestyle. There are a number of diabetes medications with varying modes of action:

- Biguanides, for example metformin, work by increasing the efficacy of insulin-regulated glucose uptake systems and reducing the intestinal absorption of glucose.

- Glitazones (thiazolidinediones), for example pioglitazone, also increase insulin sensitivity, thus increase the efficacy of insulin.

- Sulfonylureas, for example gliclazide, work by increasing insulin production by the pancreas.

- Glitinides, for example repaglinide, also increase the pancreatic output of insulin.

- Incretin mimetics (GLP-1 analogues), for example exenatide, increase the concentration of incretins—hormones linked to increased insulin production and reduced hepatic glucose output.

- Gliptins (DPP-4 inhibitors), for example sitagliptin, inhibit the DPP-4 enzyme responsible for the breakdown of incretins.

- α-Glucosidase inhibitors, for example acarbose, are also enzyme inhibitors but act by slowing down the breakdown of dietary carbohydrates to glucose.

We describe the modes of action of the different classes of medication available for lowering blood glucose levels through their physiological actions explaining the therapeutic outcome. However, physiological action is largely dependent on the chemistry of the therapeutic molecule, which is central to its physiological and therefore therapeutic action. Only when this interaction at the molecular level between the therapeutic-chemical and its biological target is fully known can the full mode of action be understood.

 The **Pharmaceutical Chemistry** book within this series explains the chemistry of therapeutic molecules.

The **Pharmaceutics** book within this series explains how drugs are taken by patients in the form of medicines. The particular dosage form in which the therapeutic molecule is presented has profound effects on the therapeutic outcome. Insulin, for example, can currently only be administered as an injection, owing to the digestion of the protein in the gastrointestinal tract. There is, however, considerable research in the area of formulating solid dosage forms and subcutaneous implants of insulin for continuous release without compromising the doses and therapeutic effects achieved.

 The study of dosage forms is covered in the **Pharmaceutics** book within this series.

In the UK there are 2.9 million people known to have a diagnosis of diabetes (Diabetes Prevalence 2011 Report http://www.diabetes.org.uk/Professionals/Publications-reports-and-resources/Reports-statistics-and-case-studies/Reports/Diabetes-prevalence-2011-Oct-2011). Diabetes thus presents a big public health issue and pharmacists have a central role in improving public health. They provide invaluable advice on all aspects of disease management, from beneficial lifestyle changes to information on the various medications available. Most pharmacies also offer diabetes testing services and help raise public awareness about the disorder.

 Public health and other professional topics are the focus of the **Pharmacy Practice** book within this series.

Pharmacists therefore have a unique position in the health care service. Their scientific background, integrated with evidence-based practice and developed professional skills and competencies, are pivotal to providing the best care for their patients.

➤ This is just the start of your challenging journey through the subject of pharmacy. The key to understanding the different approaches and topics you will encounter is to always look for and recognize the links within and between the scientific and practice elements of the profession.

FURTHER READING

Barber, J. and Rostron, C. *Pharmaceutical Chemistry,* Integrated Fundamentals of Pharmacy. Oxford University Press, 2013.

> In this book you will find a range of related topics with a focus on the chemical aspects of therapeutic molecules and their properties.

Denton, P. and Rostron, C. *Pharmaceutics: The Science of Medicine Design,* Integrated Fundamentals of Pharmacy. Oxford University Press, 2013.

> This book introduces the science of dosage forms and pharmaceutics, and explores the chemical, therapeutic, and practice implications.

Hall, J. and Rostron, C. *Pharmacy Practice,* Integrated Fundamentals of Pharmacy. Oxford University Press, 2013.

> The book outlines the main areas within pharmacy practice and introduces the concepts of the profession.

Glossary

ACE inhibitor Drug that inhibits angiotensin-converting enzyme (ACE).

Acetylcholine The neurotransmitter at preganglionic autonomic neurons and postganglionic parasympathetic neurons and at the neuromuscular junction. It acts on nicotinic and muscarinic receptors. Unusually, acetylcholine is also released from sympathetic nerves that innervate sweat glands.

Actin Thin filamentous proteins that bind to and overlap with myosin when a muscle cell contracts.

Action potential Depolarization of a cell membrane which results in the transmission of an electrical impulse.

Activation energy Energy that must be overcome in order for a reaction to occur.

Adenosine-5′-diphosphate (ADP) Nucleotide that can be converted to adenosine-5′-triphosphate (ATP). It is stored in platelets and released when platelets are activated and thus propagates the activation to neighbouring platelets. It is also an agonist for some receptors.

Adenosine-5′-triphosphate (ATP) High-energy phosphate molecule that is a product of respiration and a source of energy for cellular processes.

Adipocytes Cells that store energy in the form of fat; also referred to as lipocytes and fat cells.

Adipose tissue Connective tissue that acts as the main storage site for fat.

Adrenaline Hormone released from the adrenal medulla that acts on α- and β-adrenoceptors. The hormone is associated with stress ('fight or flight') and mediates actions which prepare the body for physical exertion.

Adrenal medulla Central portion of the adrenal gland, which produces and releases adrenaline.

α-adrenoceptors A subset of transmembrane G-protein-coupled receptors which respond to adrenaline. These receptors usually cause contraction of smooth muscle.

Afferent Neurons taking nerve impulses towards the central nervous system.

Affinity The predisposition of a ligand to bind to its receptor.

Agglutination The clumping together of molecules that occurs when incompatible blood is mixed together.

Agonist Molecule or drug that binds to a receptor (has affinity) and activates it (has efficacy). Full agonists have high efficacy, partial agonists low efficacy, and inverse agonists negative efficacy.

Albumin Main protein found in blood plasma. Its primary function is to regulate the osmotic pressure of blood.

Aldose Sugar containing one aldehyde (CH = O) group per molecule. Note that glucose 6-phosphate is an aldose in the cyclic form.

Allergen Substance that can provoke an allergic reaction.

Allosteric A type of receptor binding that involves a change in the shape and activity of an enzyme when a drug binds at a site other than the binding site of the endogenous ligand.

Amines Basic organic compounds that contain nitrogen.

Amino acid Molecule that has an amine group, a carboxylic acid group and a side-chain. Amino acids are essential for life as they are the building blocks of protein.

Amoeba Single-celled organism.

Amylase Enzyme that breaks down starch and glycogen to dextran.

Analgesic A drug that reduces the sensation of pain—a painkiller.

Anaphylactic shock A potentially lethal allergic reaction that affects the whole body.

Angina Heart condition that develops owing to lack of blood (and thus oxygen) being delivered to the heart.

Angiotensin-converting enzyme (ACE) The enzyme that converts angiotensin I to angiotensin II, which is very active in raising blood pressure.

Animals Major group of eukaryotic multicellular organisms.

Antagonist Receptor ligand or drug that does not cause a biological response itself upon binding to a receptor, but blocks or dampens agonist-mediated responses.

Anterior An anatomical term used to describe nearer the forward end of an organism.

Antibodies Singular: antibody. Also known as immunoglobulins, antibodies are Y-shaped proteins produced by B cells. They are used in the immune system to identify and destroy foreign matter such as bacteria and viruses.

Anticoagulant Substance that prevents blood coagulation.

Antigen Foreign molecule that induces an immune response (production of antibodies) when introduced into the body.

Antimuscarinic drug Drug that inhibits acetylcholine accessing the effector organs normally stimulated by the parasympathetic nervous system.

Antiporter Membrane protein involved in transporting molecules across the membrane.

Argentaffin cells Cells found in the intestine that are a rich source of serotonin.

Atherosclerosis Medical condition in which the walls of arteries thicken owing to the accumulation of fatty substances such as cholesterol.

Atrium The heart is divided into four chambers; the top two chambers are the left and right atria.

Autoimmune disorder Inappropriate response of the body's immune system against endogenous molecules and/or healthy tissue.

Autophosphorylation Addition of phosphate groups to a protein catalysed by its own enzymic activity.

Autosomal Any chromosome that is not a sex (X or Y) chromosome.

Baroreceptors Receptors that are sensitive to a change in pressure.

Basophil Polymorphonuclear leucocyte which has an affinity for high-pH stains.

β globin Polypeptide that forms part of the haemoglobin protein in the blood.

β-lactam antibiotics Broad class of antibiotics, including penicillins and cephalosporins, that have a β-lactam ring at the heart of their structure. They interfere with bacterial peptidoglycan cell wall synthesis.

Biconcave Concave on both sides or surfaces.

Bile salts Aid the digestion of dietary fats and the removal of toxins. They are produced by the liver and stored in the gall bladder.

Blood–brain barrier Notional barrier that prevents highly charged chemicals from accessing the brain.

Body systems Group of organs that function together to achieve a common outcome.

Bradycardia Slowing of the heart rate.

Bronchoconstrictor Substance that induces the narrowing of the bronchi.

Brown adipose tissue Specialized type of fat tissue that produces heat from fat metabolism. It is found between the shoulders.

Cancer A number of diseases that are characterized by abnormal growths and uncontrolled cell division and often have the ability to spread (metastasize) to other tissues in the body.

Carcinogenic Cancer-causing.

Cardiac output Volume of blood pumped by the heart in 1 minute.

Cardiomyocyte Heart (cardiac) muscle cell.

Catecholamines Adrenaline, noradrenaline, and dopamine can all be described as catecholamines because their chemical structure includes an amine group and a catechol ring.

Cerebrospinal fluid Liquid that fills the cavities of the brain and the spinal cord. Somewhat similar to blood plasma.

Cervical Part of the neck—the first seven vertebrae of the spinal cord.

Chemoreceptors Receptors that respond to chemicals binding to them, such as agonist drugs.

Chemotactic agents Chemicals that are attractive to certain cells of the body's defensive system.

Chemotaxis Process by which cells direct their movement according to the presence of certain chemicals in their environment.

Chemotherapy Treatment of cancer with chemicals/drugs that kill rapidly dividing cancerous cells in the body.

Cholinergic nerves Nerves that release acetylcholine.

Cholinesterase Family of enzymes that catalyse the hydrolysis of acetylcholine into choline and acetate.

Chromaffin cells Adrenaline-producing cells of the adrenal glands, which are situated above the kidneys.

Chronotropic effect Change in the rate of activity.

Clinical trials Generally considered to be biomedical or health-related research studies in humans that follow a predefined method.

Coding strand The strand in a double-stranded DNA molecule that has the same sequence of bases as the transcribed mRNA (except that uracil (U) in RNA replaces thymine (T) in the DNA sequence). The other strand in the double-stranded DNA molecule is often referred to as the non-coding strand.

Codominant Where the characteristics of genes on both chromosomes are expressed equally and neither gene is dominant over the other.

Coenzyme Accessory factor that is required for an enzyme to function effectively.

Collagen Protein predominant in bone and connective tissues.

Connective tissue Tissue type that functions to support and connect other types of tissues and organs. It is composed of cells, fibres, and an extracellular matrix and is found throughout the body. Bone, cartilage, and tendons are structures formed from connective tissue.

Coronary artery disease Disease of the heart in which atherosclerotic plaques obstruct normal blood flow.

Corpuscle The endings of sensory nerves that can detect changes in the environment and activate sensory neurons.

Corticosteroids Steroid hormones produced in the cortex (outer part) of the adrenal gland and which are involved in a variety of physiological processes.

Covalent bond Strong form of chemical bonding that occurs when two different atoms share electrons.

Cranial Of or relating to the skull/cranium. Cranial nerves are nerves that emerge directly from the brain rather than the spinal cord.

3′,5′-Cyclic adenosine monophosphate (cyclic AMP, cAMP) Derivative of ATP that acts as a second messenger in signal transduction within biological processes. Contains a ribose sugar with an adenine base attached to the 1′ carbon atom of the sugar and a phosphate attached at both the 3′ and 5′ carbon atoms, giving a cyclic structure.

Cysteine Amino acid containing a thiol (–SH) group.

Cytokines Chemicals, usually protein in nature, that are secreted by some cells as signalling molecules that help with intercellular communication.

Cytoplasm Thick, aqueous liquid that anchors the organelles within in the cell.

Daughter cells The two cells that result when one cell divides (cell division).

Deiodination Removal of iodine.

Denaturation Irreversible damage to a protein structure usually by heat.

Deoxyribonucleic acid (DNA) Type of nucleic acid that contains the genetic instructions for all development and functioning of most living cells.

2′-Deoxyribose The pentose sugar ribose with an oxygen atom missing at the 2′ position ($C_5H_{10}O_4$).

Depolarization Process whereby an excitable cell moves from having a negative resting potential to a positively charged membrane.

Differentiate Process by which a cell such as a haematopoietic stem cell becomes a more specialized cell such as a red blood cell.

Dimerization Chemical reaction whereby two identical molecules are joined into a dimer.

Dopamine Neurotransmitter molecule.

Dosage regimen The formulation, doses, and schedule of administering a drug to maintain a steady concentration in the blood plasma.

Downstream Toward the 3′ end of a DNA molecule relative to a particular position in the molecule.

Drug target Molecule or structure within an organism, linked to a particular disease, which may be blocked (inhibited) or activated by a drug.

Duodenum The first organ of the small intestine.

Dyspepsia A disturbance of digestion where there is excess acid in the stomach.

EC_{50}/ED_{50} Dose/concentration of drug that produces a half-maximal response.

Efferent Neurons taking nerve impulses away from the central nervous system.

Efficacy Ability of the receptor–ligand complex to bring about the maximal physiological response. Also called intrinsic activity.

Electrochemical gradient Tendency of an ion to diffuse owing to both its concentration difference and its electrical potential relative to the membrane potential.

Electrostatic bonds/interactions Chemical bonds in which one atom loses an electron and another gains an electron.

Electrostatic interaction The attraction between a charged group on one molecule and an oppositely charged group on another molecule.

Enantiomers Chemical term describing how two molecules with the same atoms are mirror images of each other.

Endocrine gland Organ that secretes substances directly into the blood circulatory system.

Endocytosis Process by which substances gain entry to the cell without passing through the cell membrane, involving an infolding and engulfing of the cell membrane around the cell to form an 'endosome'.

Endogenous Originating from or produced by an organism.

Endometrium Inner membrane of the uterus.

Endothelial Endothelium is a thin layer of endothelial cells that lines the interior (inside) of blood vessels.

Endothermic Reaction that requires heat.

Enzyme Protein catalyst that increases the rate of biochemical reactions in the body.

Epithelial cells Group of cells that are often arranged in layers that cover our external and internal surfaces to form the epithelium. The cells that make up our skin are epithelial cells.

Equilibrium State in which the concentrations of the substrates and products have no net change.

Equimolar concentrations Concentrations that are of the same molarity.

Exocytosis A process whereby substances contained within vesicles in the cytoplasm are secreted from cells by the fusion of the vesicles with the cell membrane.

Feedback mechanism The outcome of the process regulates the same process.

Femtolitre One thousand million million (one quadrillionth) of a litre (using the short scale large-number naming system), 1×10^{-15}.

Follicular cells Cells that surround a cavity.

Fungi Subgroup of eukaryotic organisms, including moulds, yeasts, and mushrooms, that have a cell wall containing a substance called chitin.

Gap junction Connection between two adjacent cells connecting the cytoplasm of each cell to allow the transfer of molecules between them.

Gene A sequence of DNA that codes for a functional substance such as a protein or for one of the different forms of RNA molecule.

Gene promoter A segment of DNA that usually occurs upstream from a gene coding region and acts as a controlling element in the expression of that gene.

Gene transcription The production of RNA from a DNA template, usually prior to protein synthesis.

Genetic code The code describing the translation of the sequence of nucleotides found in nucleic acids into a sequence of amino acids. It is written in a series of base sequence triplets (codons) of the nucleotides. There are 64 possible codons in the genetic code, which can act as signals or code for amino acids.

Genus Level of taxonomic classification of living and fossilized organisms.

Germ cells An egg or sperm cell. These cells have half the normal number of chromosomes but the full complement of chromosomes is restored at the point of conception, when two germ cells unite.

Gestational diabetes Type of diabetes that develops in some pregnant women.

Gland Organ that stores and releases biological substances.

Glucagon Peptide hormone released from the pancreas. It causes the liver to release stored glucose into the bloodstream.

Gluconeogenesis The creation of glucose from a non-carbohydrate source.

Glycation The attachment of glucose to other substances.

Glycogen Complex branched carbohydrate molecule consisting of glucose units. Used as storage of energy in the cell.

Glycoside A chemical compound with a bound sugar molecule.

G-protein-coupled receptors Receptors found in the cellular membrane and associated with a G-protein. Involved in activating intracellular signalling pathways.

Growth factor Substance that effects the growth of an organism or organ by stimulating cell proliferation.

Haemodialysis Process of removing waste products from the patient's blood.

Haemodynamic pressure The pressure of the blood induced by the pumping of the heart.

Haemolysis Rupture of red blood cells. Usually follows an incompatible blood transfusion.

Heparin Naturally occurring inhibitor of blood clotting (anticoagulant) found in mast cell granules and usually released in response to injury.

Hepatocyte   Liver cell.

Histamine   Histamine is produced by basophils and mast cells. It is released into the body as part of the immune response to invading pathogens. It is also produced as a regulator of digestive function.

Homeostasis   The body's control of its internal environment.

Hormones   Chemical messengers released by endocrine glands and disseminated around the body by the bloodstream.

Host cell   Cell that harbours a foreign entity, for example a virus, bacteria or extrinsic DNA or proteins.

Humoral transmitters   Chemicals that are released from nerves and convey information to other cells.

Hydrogen bond   Form of electrostatic interaction specifically mediated by the attraction between a hydrogen atom within one molecule and a negatively charged atom within another molecule.

Hydrogen bond acceptor   An electronegative atom that 'accepts' a hydrogen atom which is covalently bonded to another electronegative atom.

Hydrogen bond donor   Electronegative atom such as fluorine, oxygen, or nitrogen which is covalently bonded to the hydrogen atom that is 'donated' to another electronegative atom.

Hydrolysis   The chemical process whereby a water molecule is added to a substance, causing that substance to be split into two parts. One part of the initial substance gains a hydrogen ion (H^+) and the other gains a hydroxyl group (OH^-) from the water molecule.

Hydrophilic   Water attracting.

Hydrophobic   Water repelling.

Hydrophobic bond   Attractive force between non-hydrophilic (water-repelling) parts of two molecules.

Hypertension   Pathological elevation in blood pressure.

Hypothalamus   Region of the brain involved in the control of body temperature and other homeostatic mechanisms.

Immune system   Mechanism by which the body protects itself from pathogens such as bacteria, viruses, fungi and parasites.

Indole ring   Two-ring heterocyclic structure found in the amino acid tryptophan.

Inotropic drug   A drug that increases the strength of heartbeats.

Inotropic effect   Change in the force of an activity.

Insulin   Peptide hormone released from the pancreas. It causes glucose to be taken up from the blood into tissues, where it is used or stored.

Interleukins   Group of proteins that are involved in communication between leukocytes and have a role in the inflammatory response.

Intranasal   Drug delivery via the nose.

In vitro   Conducted or performed outside of living organisms (Latin: in glass). The opposite is *in vivo* (Latin: within the living), which refers to experiments conducted within living organisms.

Ion channel   Protein complexes spanning the cellular membrane, which when bound by an appropriate ligand undergo a conformational change creating a pore or channel through which ions can enter the cell.

Ionic bond   Chemical bond formed by an electrostatic attraction between two ions with opposite charges.

Islets of Langerhans   Areas of the pancreas responsible for the release of hormones controlling blood sugar levels.

Isomerization   Reaction in which a molecule is transformed into a different molecule with the same atoms but rearranged. The product is called an isomer.

Ketose   A sugar containing one ketone group per molecule. Note that fructose 6-phosphate is a ketose in the cyclic form.

Lateral horn   Horns from the spinal cord composed of visceral neurons.

Laws of thermodynamics   Four laws defining physical quantities, including energy transfer, entropy, and temperature.

Leukaemia   A cancer of precursor white blood cells (leukocytes).

Ligand   Molecule or ion that binds to a receptor and affects its function.

Ligand-gated ion channel   Receptors that are ion channels, such that when a ligand binds the ion channel opens.

Lipids   Biomacromolecules, including fats, which function within the body as energy storage, structural elements within the cellular membrane, and signalling molecules.

Lipophilic   Having an affinity to lipids.

Liposomes   Microscopic artificial lipid bilayer spheres that enclose an aqueous core and are sometimes used to deliver drugs or other substances to cells.

L-type Ca^{2+} channel   A type of voltage-gated ion (calcium) channel.

Lumbar   Five vertebrae of the lower spine above the sacrum.

Lymphatic system   Circulatory system within the body used to distribute lymph, consisting of lymphocytes and plasma.

Lymphoid   One of the pathways of haematopoiesis, in which the white blood cells called lymphocytes are formed.

Macrophages   White blood cells found in the tissues and formed from monocytes. They remove pathogens and cell particles within the tissue by a process known as phagocytosis.

Malignant   Tendency of a condition to become progressively worse. A malignant tumour denotes cancer.

Mammals   Class of endothermic vertebrate animals that includes humans. Characterized by hair or fur on the skin and lactation as the means by which the female feeds her young.

Mast cells   Cells found in epithelial tissues. They are derived from basophilic polymorphonuclear leucocytes and are rich in histamine.

Mechanoreceptors   Receptors that respond to a mechanical change or movement.

Membrane potential (resting)   Excitable cells actively maintain ionic gradients to keep their membranes negatively charged—this is considered the resting membrane potential.

Mesentery Thin layer of connective tissue that holds the gastrointestinal tract in place within the abdominal cavity.

Messenger signals Endogenous molecules produced by the body to carry a message from one cell to another.

Metabolism Range of chemical reactions, vital for life, in which substances are produced or broken down to produce energy.

Metabolites Intermediary molecules arising due to the metabolism of another molecule.

Methylated Containing a methyl group.

Microtubules Hollow cylindrical proteins, found in the cytosol of eukaryotic cells, which give structural support to cells.

Microvascular integrity The state of small blood vessels.

Micturition Urination.

Mineralocorticoids Group of corticosteroids involved in the regulation of sodium and potassium ion movement across the cell membrane.

Mitochondria Organelles found within the cytoplasm of a cell. They create and provide the cell with energy.

Mitochondrial electron transport chain The pathway in which electron transfer is coupled to proton (H^+) transfer across a membrane. This results in an electrochemical proton gradient, which is used in the mitochondria to generate energy in the form of ATP.

Molar Unit of concentration meaning the number of moles per unit volume, for example mol/litre.

Mole Unit of measurement to describe the amount of a chemical. One mole of an element contains 6.02×10^{23} atoms of that element. One mole of a molecule contains 6.02×10^{23} formula units of that molecule. 6.02×10^{23} is Avogadro's constant.

Monoamine neurotransmitters Neurotransmitters that have a similar chemical structure that includes one amine group connected by a two-carbon chain to an aromatic ring. Examples include dopamine, serotonin, and noradrenaline.

Monoclonal antibodies Antibodies that are all exactly the same because they are made by identical B cells.

Motor endplate Junction between a motor neuron and a muscle.

Muscarinic receptors Receptors that can be stimulated by muscarine and acetylcholine. Usually found on organs stimulated by the parasympathetic nervous system.

Myelinated Coated in the insulating myelin sheath produced by neighbouring glial cells.

Myeloid Originating in the bone marrow and used to describe a white blood cell that is not a lymphocyte.

Myocyte Muscle cell. Cardiac myocytes are muscle cells in the heart (sometimes called muscle fibres).

Myosin Thick filamentous protein that binds to and overlaps with actin when a muscle cell contracts.

Natriuretic peptides Class of peptide hormones that cause vasodilatation in many blood vessels and increased sodium and water loss from the kidneys.

Negative feedback system Mechanism by which the product of a system inhibits its own release.

Nephron Part of the kidney that carries out its main functions of filtering the blood and excreting urine.

Neurotoxin Poisonous chemical that acts on nervous tissue to interfere with nerve function.

Neurotransmitter Chemical messenger that carries a signal from one neuron to another.

Neutrophils Polymorphonuclear leucocytes that have an affinity for neutral pH stains.

Nicotinic acetylcholine receptor (nAChR) Cholinergic receptor that is an ion channel, which binds and responds to acetylcholine and nicotine (but, unlike muscarinic receptors, does not respond to muscarine).

Nitrate-like effect Stimulation of guanylate cyclase that causes an increase in the level of cyclic guanosine monophosphate, which in turn activates protein kinase G and initiates vasodilatation.

Nociceptive Describes receptors, nerve fibres, or pathways that respond to noxious stimuli, such as mechanical, chemical, or thermal irritants, and elicit pain.

Non-covalent bonds Weaker than covalent bonds and include electrostatic interactions, hydrogen bonds, hydrophobic interactions, and van der Waals interactions.

Non-essential amino acid Amino acid that can be made by the body.

Non-specific side effects Side effects that cannot be attributed to the specific pharmacological action of a drug.

Non-steroidal anti-inflammatory drugs (NSAIDs) Drugs such as aspirin and ibuprofen, which are not steroids and can reduce pain, fever, and inflammation.

Noradrenaline A neurotransmitter with a similar chemical structure to adrenaline. It is released from sympathetic nerves and activates α-adrenoceptors.

Nucleotides Molecules comprised of a nitrogenous base, a five-carbon sugar unit, and phosphate groups.

Nucleus Organelle containing the genetic material of a cell.

Organelles Structures within the cytoplasm of a cell that carry out specialized functions.

Osmolarity Measure of solute concentration.

Osmosis Movement of water from an area of high solute concentration to an area of low solute concentration through a semipermeable membrane.

Osmotic pressure Pressure required to stop the tendency of a solvent (usually water) from passing from one area to another through a semipermeable membrane (via osmosis).

Oxidation–reduction Reactions where the oxidation state of atoms are changed. In oxidation an atom of oxygen is gained or a proton/hydrogen removed or an electron is lost; in reduction an atom of oxygen is removed or a proton/hydrogen is gained or an electron is gained.

Parasympathetic nerves Occur in the autonomic nervous system and are typically responsible for 'rest and digest' signals.

Parathyroid gland Endocrine gland involved in calcium homeostasis. It is situated close to the thyroid gland.

Parietal cells Cells that secrete hydrochloric acid into the stomach.

Pathologies Disease states. Sometimes used more selectively to refer to disease processes at the molecular and cellular level.

Penicillin Antibiotics derived from the fungus *Penicillium*.

Peptic ulceration An erosion of the lining of the oesophagus, stomach, or duodenum.

Peptide bond A covalent bond formed between a carboxyl group and an amino group.

Peristalsis Rhythmic contraction of the intestine that moves the contents along its length.

Pharmacodynamics Study of the biochemical and physiological effects therapeutic agents have on the body.

Pharmacogenomics Study of the influence of genetic variation on the response of patients to drugs by correlating gene expression with the physiological effect of the drug.

Pharmacokinetics Study concerning the outcome (absorption, distribution, metabolism, and excretion) of therapeutic agents administered to living organisms.

Pharmacology Study of drugs, their effects on the body, and their clinical uses.

Phenotype Observed inherited trait, such as blood groups.

Phosphoinositides Building blocks of cell membranes.

Phosphorylation Process of adding a phosphate group (PO_4^{3-}) to a protein substrate.

Plant Multicellular eukaryotic organism that can carry out photosynthesis.

Plasmin Enzyme that breaks down the fibrin meshwork of blood clots during the process of fibrinolysis. Plasmin is the active form of the protein plasminogen, which is produced in the liver.

Platelets Blood cells responsible for coagulation.

Polymers Compounds formed of many repeated simpler molecules (monomers).

Polypeptide A chain of amino acids (between 10 and 100) that are linked by peptide bonds.

Positive feedback mechanism Mechanism by which the product of a system increases its own output.

Posterior An anatomical term used to describe the back or rear end of an organism.

Potency The amount (concentration) of a drug required to produce an effect of a given intensity.

Presynaptic nerve/neuron Afferent nerve that carries an electrical signal to a synapse (nerve junction) and releases a neurotransmitter into the synaptic cleft.

Proliferative disease Disease in which excessive cell division (cell proliferation) occurs.

Proprioception Awareness of muscle and joint movement in relation to other body parts.

Prostaglandin Biologically active long-chain polyunsaturated fatty acid with an internal five-carbon ring.

Prosthetic group Coenzyme (small organic cofactor) tightly bound to a protein.

Protease Enzyme that catalyses the breakdown of peptide bonds in proteins.

Protease inhibitor Substance that inhibits enzymes which break down proteins.

Proteins Biomacromolecules consisting of amino acids. Their sequence is defined by the nucleotide sequence within the relevant gene.

Protists Class of eukaryotic organisms that include protozoa, algae, and some slime moulds.

Proton-pump inhibitors Drugs that reduce acid levels in the stomach by inhibiting the secretion of hydrogen ions by parietal cells.

Purine Molecules that consist of a six-membered ring (two nitrogen and four carbon atoms) fused to a five-membered ring (two nitrogen and three carbon atoms). Purines include the bases adenine (A) and guanine (G).

Pyrimidine Molecules that consist of a single six-membered ring made from two nitrogen and four carbon atoms. Pyrimidines include the bases cytosine (C), thymine (T), and uracil (U).

Pyrogens Substances that can induce fever.

Radiotherapy Treatment of cancer and other diseases with ionizing radiation.

Rami communicantes Bridge-like structures between the spinal nerves and the paravertebral ganglia known as the sympathetic trunk (otherwise referred to as the sympathetic chain or gangliated cord), which travels from the base of the skull to the coccyx.

Receptor A protein on the cell surface or within a cell that receives and responds to a molecule (ligand), which can be an endogenous ligand or a drug molecule.

Recombinant human insulin Insulin produced by bacteria that have been transfected with human DNA.

Red bone marrow Tissue found inside bones that is the site of erythropoiesis.

Reflex Produced when sensory neurons send information to the spinal cord which then directly synapses onto autonomic efferents to rapidly produce an effect at the target organ without the involvement of the higher centres of the nervous system (the brain).

Repolarization Process whereby the membrane of a depolarized cell returns to the negative resting potential. This is achieved by the movement of K^+ ions out of the cell.

Respiration The cellular process by which glucose is transformed to NADH and ATP.

Ribonucleic acid (RNA) An important constituent of all living cells that encodes genetic information.

Ribose Pentose sugar found in RNA nucleotides ($C_5H_{10}O_5$).

Sacrum The five lowest bones of the vertebral column, which normally fuse in adulthood.

Sarcoplasm Intracellular fluid of a muscle cell. It is similar to the cytoplasm of other cells but is especially rich in energy stores and oxygen-carrying molecules. 'Sarco' prefixes some elements of cells when describing muscle cells.

Sarcoplasmic reticulum Cellular organelle that is a type of smooth endoplasmic reticulum found in muscle cells. Sarcoplasmic reticulum contains large stores of calcium ions, which it releases when the muscle cell is stimulated to contract.

Saturated   Molecule containing no double or triple bonds.

Second messenger   Molecule that passes a signal from a membrane receptor to its target molecules within a cell.

Serotonin   Monoamine neurotransmitter found in the brain and intestines. It is sometimes referred to by its chemical name, 5-hydroxytryptamine (5-HT).

Severe combined immunodeficiency disease (SCID)   Disease in which children are born with a non-functioning immune system, making them extremely vulnerable to severe infections.

Somatosensory system   The biological structures that detect and process sensory information.

Steric clash   When atoms within molecules collide with each other due to lack of space.

Stroke volume   The volume of blood pumped by one ventricle of the heart in a single beat.

Subcellular   Located or occurring within a cell, for example an organelle or cell component.

Subcutaneous   Drug delivery by injection under the skin.

Superior vena cava   Vein that carries deoxygenated blood from the upper half of the body to the heart's right atrium.

Superoxide radical   Chemical with the formula O_2^-.

Sympathetic nerves   Occur in the autonomic nervous system and are typically responsible for 'fight or flight' signals.

Synapse   Structure between a neuron and another cell that conveys the electrical and/or chemical signals.

Synaptic cleft   The space between a presynaptic neuron and a postsynaptic neuron.

Tachycardia   A faster than normal heart rate.

Teratogenic   Causing malformations to a foetus.

Tetramer   Protein with four subunits.

Thoracic   Pertaining to the chest. The thoracic cavity is the chest and the thoracic vertebrae are the 12 vertebrae between the cervical and lumbar vertebrae.

Thrombosis   Blood clot formed inside a blood vessel.

Thymus   Specialized organ of the immune system. It is involved with the maturation of T cells.

Tissues   Collection of similar cells and other substances that work together to carry out specific bodily functions.

Transcription factor   Protein that binds to DNA and regulates the transcription of the genetic information.

Transdermal   Drug delivery across the skin (for example a nicotine patch).

Translation   Synthesis of a polypeptide chain using messenger RNA as a template.

Triacylglycerol   Major energy reserve found in animals, which is an ester derived from glycerol with three identical fatty acids.

Tumour necrosis factor-α   Cytokine involved in many cellular interactions involving the body's defensive systems.

Tyrosine kinase receptor   Receptors found in the cellular membrane which have a tyrosine kinase domain able to phosphorylate tyrosine residues of proteins.

Unsaturated   Molecule that possesses at least one double or triple bond. The degree of unsaturation dictates the quantity of hydrogen ions that can bind to the substance.

Upstream   Toward the 5′ end of a DNA molecule relative to a particular position in the DNA molecule.

Urticaria   Allergic, itchy skin rash.

Vaccine   Medicine that stimulates the immune system to produce antibodies to fight a specific disease.

Vacuole   Organelle found in the cytoplasm of a cell, surrounded by a phospholipid membrane and often used for storage.

Vagus nerve   Also known as cranial nerve X, most of the innervation of parasympathetic nervous system uses this nerve.

Van der Waals interaction   Non-covalent bond that develops when the electronic charge around an atom is not equally distributed and induces an opposing asymmetric electron distribution in another molecule.

Varicosities   Enlargements or swellings of the blood vessels.

Vasoconstriction   Constriction or narrowing of the blood vessels.

Vasodilatation   Widening of blood vessels mediated by the relaxation of muscle cells contained within the blood vessel wall.

Vasopressin   Hormone released from the hypothalamus that promotes water reabsorption in the kidneys.

Ventral spinal root   Nerves that leave the vertebral column.

Ventricle   The heart is divided into four chambers; the lower two chambers are the left and right ventricles.

Vesicles   Bubble-like sacs surrounded by a lipid layer, used by the cell to transport cellular products.

Viruses   Simple infectious agents that are parasitic to living cells and can only multiply within them. Many viruses are harmless but some are responsible for a range of diseases such as colds, influenza, and some cancers.

Visceral   Relating to the viscera (the internal organs).

von Willebrand factor   Glycoprotein found on the plasma membrane surface of platelets.

Answers to self check questions

Chapter 1

1.1 A molecule must be able to interact with the living organism to be able to adjust the processes that are naturally occurring and consequently cause a change that is clinically beneficial for the patient for it to be considered a therapeutic agent (drug).

1.2 It is important to establish if there are any hereditary disorders that may have been passed on to the patient. This may help diagnose the ailment and decide on the appropriate treatment.

1.3 Metabolism covers all the reactions in the body involved in harvesting energy from nutrients and using the energy to build complex molecules. Our bodies use energy not only to conduct physical activities but also to facilitate vital bodily functions required to sustain life.

1.4 Homeostasis encompasses the principle that there are a number of complex processes simultaneously occurring, and the body maintains them in balance so it is functioning within normal range, close to the 'set-points'.

1.5 Receptors are proteins, either within the cellular membrane or within the cell (intracellular), which recognize a particular chemical signal and initiate a specific cellular response that results in a physiological effect.

1.6 The main difference is the location of the target cell in relation to its secreting cell body. Hormones are secreted by endocrine glands and are required to travel via the blood from one part of the body to the target organs. Neurotransmitters are secreted by nerve cell endings and these are close to the target cell; however, the nerve cell body could be some distance away from the terminal.

1.7

1. Act upon cellular receptors, either stimulating the physiological response (agonists) or preventing the response from occurring (antagonists).

2. Affect enzyme reactions by inhibiting the enzyme, increasing the concentration of the enzyme, or acting as a substrate for the enzyme reaction.

3. Change the environment of cells, tissues, and organs.

4. Alter the expression of genes.

5. Eliminate infecting organisms by altering the parasite's cellular processes.

1.8 A drug taken orally is absorbed through the epithelium of the gastrointestinal (GI) tract. The blood from the GI tract passes through the liver, where a proportion of the absorbed drug is metabolized before it is systemically distributed. Drugs that are administered parenterally have 100% bioavailability and are distributed around the body before their metabolism begins. See Figure 1.7.

1.9 The therapeutic window refers to the dose range within which there is sufficient drug to cause physiological change without an increased risk of side effects arising or toxic effects occurring. A dose higher than the recommended therapeutic window may have significant toxic and adverse effects, while a dose below the therapeutic window will be ineffective and not produce sufficient therapeutic effects. For a drug with a narrow therapeutic window it will be harder to achieve an appropriate dose within the patient and the risks of over-dosing or under-dosing are greater.

Chapter 2

2.1 Eukaryotes have a membrane-bound nucleus and other specialized and defined structures called organelles within their cytoplasm whereas prokaryotes do not.

2.2 The complementary sequence of the other strand is 5'-ACTCGTAAT-3'.

2.3 The three main differences between DNA and RNA molecules are:

1. The sugar 2'-deoxyribose is used in DNA whereas ribose is used in RNA.

2. DNA is usually double stranded whereas RNA is usually single stranded.

3. In DNA, the base thymine (T) is used whereas in RNA the base uracil (U) is used.

2.4 See Table on next page.

2.5 DNA polymerase is only able to synthesize a new strand of DNA in the $5' \rightarrow 3'$ direction. The leading template strand runs in the $3' \rightarrow 5'$ direction and so its new complementary strand has a $5' \rightarrow 3'$ direction that can be synthesized in one continuous run, following the replication fork as it 'unzips' the DNA. However, the polymerase also has to copy the lagging template strand, which runs in the $5' \rightarrow 3'$ direction, and synthesize a new DNA lagging strand in the opposite direction, running away from the replication fork. Again, this is because the polymerase can only synthesize DNA in the $5' \rightarrow 3'$ direction. In order to follow the replication fork as it opens the DNA, the polymerase has to synthesize the new lagging strand in a discontinuous fashion, creating as a series of short sequences called Okazaki fragments.

2.6 This is because men have one X and one Y chromosome and so will always express the X-chromosome linked mutated genes if present and thus have the disease. Females, on the other hand, have two X chromosomes so, in principle, the normal gene on one chromosome should compensate for or override the mutated gene on the other. However, all females effectively only have one X chromosome because one of the pair is switched off by a process

						Start			
-	-	-	-	-	-	Methionine	Aspartic acid	Glutamine	Glycine
G	UAU	CCG	AGA	GAG	GAC	AUG	GAU	CAG	GGU

Cysteine	Glutamic acid	Glycine	Aspartic acid	Tyrosine	Lysine	Arginine	Tryptophan	Cysteine	Aspartic acid
UGU	GAA	GGA	GAU	UAC	AAA	AGA	UGG	UGU	GAU

						Stop			
Isoleucine	Glutamic acid	Tyrosine	Leucine	Glycine	Glutamic acid	Asparagine	-	-	-
AUA	GAG	UAU	CUG	GGA	GAA	AAC	UAG	AGG	AAU

-	-	-	-	-	-	-	-	-	-
AAA	AAA	AAA	AAA	AAA	AAA	AAA	AAA	AAA	AAA

of X-inactivation during embryonic development. The chromosome that is inactivated is random but, on average, most females have about equal numbers of their cells with the paternal and maternal X chromosome inactive. In most X chromosome-linked recessive genetic diseases the cells carrying the normal gene can compensate for those carrying the mutated gene and the result is very mild or no symptoms at all. However, as X-inactivation is a random process that occurs during development occasionally and unfortunately in some females the normal X chromosome is inactivated in many more of their cells than the mutated X chromosome. This can result in a mild to severe disease state depending on the proportion of normal X chromosome inactivation.

2.7 The vaccination protects children against getting HPV but the vaccine is not a HPV treatment. This vaccine is offered to schoolgirls of this age because they are unlikely to be sexually active and so are unlikely to have already been infected with HPV.

Chapter 3

3.1 Globular proteins are spherical proteins that are highly soluble in aqueous solutions because their hydrophilic amino acids are on the outside of the molecule while the hydrophobic amino acids are on the inside. Fibrous proteins are linear proteins that are insoluble in aqueous solutions because the hydrophobic amino acids are on the outside of the molecule. Likewise, membrane proteins have hydrophobic amino acids on the outside that enable them to interact with the non-polar phase of the cell membrane.

3.2 The primary structure is stabilized by cross-links called disulphide bonds. These occur between cysteine residues, which are oxidized to form a cystine unit.

The secondary structure is stabilized by hydrogen bonds between the NH and C = O groups within the polypeptide backbone. In the α-helix, hydrogen bonds form between an NH group and the NH group that is four residues ahead. In β-strands, hydrogen bonds form between the side chains of adjacent strands. The stability of an α-helix is also achieved by excluding certain amino acids which cause instability. These include proline, isoleucine, valine, threonine, aspartate, serine, and asparagine.

The tertiary structure is stabilized by reducing the surface area to volume ratio by keeping hydrophobic and hydrophilic residues in separate regions. Hydrogen bonds again occur between NH and C = O groups within hydrophobic regions, and van der Waals interactions occur between tightly packed side chains.

The quaternary structure is stabilized by hydrogen bonds which aid the formation of extra non-covalent interactions and disulphide bonds. Again, stability is enhanced by separate hydrophobic and hydrophilic regions and the reduction of the surface area to volume ratio.

3.3 The enzyme active site is the three-dimensional cleft or crevice formed when the polypeptide chain folds. It excludes water and other non-substrate molecules and is where the substrate binds and catalytic activity occurs. The amino acids forming the active site are few in number and may be several residues apart. Some residues interact with the substrate. In enzymes with a quaternary structure, each monomer or protein chain may contain part of the catalytic site and so function is only achieved when multiple polypeptides form the complete active site.

3.4 The concentration of substrate at the start of the reaction, as close to zero as possible, is the concentration that has been added. On addition of enzyme, the rate at which product formation occurs is increased. It must be noted that the reaction equilibrium is unchanged. The rate at which the product is formed is called the velocity, V. When the substrate concentration is low, V is proportional to the substrate concentration. The concentration of enzyme is not limiting and the number of active sites available for catalytic conversion is in excess of the concentration of substrate. However, when the concentration of substrate increases, more active sites of the enzyme are occupied and become limiting. V is thus not proportionally increased. Instead, it is independent of the substrate concentration and begins to level off to form a plateau. This plateau is called the saturation point and is the point where the concentration of substrate saturates the enzyme active sites so that no further conversion can take place.

3.5 Irreversible inhibitors bind to the enzyme as a normal substrate would, but when catalysis starts an intermediate is generated. This intermediate covalently or non-covalently modifies the structure of the enzyme active site. Such inhibitors are normally specific for a type of enzyme and modify a specific residue; for example, penicillin covalently modifies the serine residues in the active site of the transpeptidase enzyme. Therefore by using a panel of known irreversible inhibitors with the ability to modify different residues it is possible to determine whether these residues are in the active site or not, by measuring whether inhibition has occurred.

3.6 A long-distance runner needs a greater amount of energy spread over a longer period of time than someone doing a short burst of activity or no activity at all. Carbohydrates are the predominant fuel used by the body. There are two forms: the simple sugars and the complex carbohydrates. The simple sugars can be metabolized directly but the complex carbohydrates are first broken down to simple sugars, namely glucose. This is vital to life as the brain and red blood cells cannot utilize fuels other than glucose. If the runner consumes a small amount of carbohydrate the glucose enters the bloodstream and is transported around the body. This is vital for the muscles when exercising. While the blood glucose level is high, the glucose is metabolized to give energy in the form of ATP. This occurs during glycolysis and the product, pyruvate, then enters the citric acid cycle to give further molecules of ATP, CO_2, and H_2O. The body also has a carbohydrate store in the form of glycogen in the liver. Once the glucose from the diet, that is the blood glucose, is depleted the glycogen is then broken down to release more glucose to feed into glycolysis.

However, if activity persists and the carbohydrate intake remains low because the runner has not stocked up, the body then begins to metabolize other fuels, including fats or lipids and then proteins. Fats in the adipose tissue are broken down to fatty acids, which then undergo β-oxidation to give acetyl coenzyme A (acetyl CoA). This then enters the citric acid cycle if there is some carbohydrate remaining. If not, the acetyl CoA is used to produce glucose by the gluconeogenic pathway, which results in the production of acetoacetate and D-3-hydroxybutyrate.

Collectively, these molecules are called ketone bodies. If the long-distance runner does not have sufficient adipose tissue for this to take place, protein in the muscle starts to be broken down. The resulting amino acids can then be further broken down into glucose in the liver. The amino group is first removed and the remaining carbon skeleton is metabolized via formation of glucose, a citric acid inter-mediate, or acetyl CoA. The 20 common amino acids are degraded by different pathways to converge at seven products. These include acetyl CoA, acetoacetyl CoA, fumarate, α-ketoglutarate, oxaloacetate, pyruvate, and succinyl CoA. These are substrates for glycolysis or the citric acid cycle and thus can be used to form ATP. Without these extra pathways the runner would die.

Chapter 4

4.1 Tyrosine kinase receptors trigger several biological pathways important in proliferation of cells and are therefore important targets for anticancer therapies. They are membrane-bound receptors that have an exofacial (facing outwards from the cell) domain that binds ligands such as growth factors and, of course, certain drugs. The exofacial domain is connected across the membrane to an endofacial domain in the cytoplasm.

When two of these receptors dimerize, or come together, they accept phosphate groups. This process, known as autophosphorylation, activates the tyrosine kinase enzymic activity of the receptor pair. This enzyme activity consists of the phosphorylation of tyrosine residues in certain proteins and holds the key to the activation of several pathways involved with the growth and proliferation of cancer cells. These pathways are often triggered by a first step in which intermediate proteins in the pathways are activated when certain tyrosine residues are phosphorylated. An example of a tyrosine kinase receptor is the epidermal growth factor receptor, the target of many novel anticancer agents.

4.2 Binding of drug to a receptor initiates the downstream response that determines its clinical effects. There are at least five classes of receptor. Some of these are true receptors, where specific binding of the endogenous ligand or a drug to a complementary site triggers a downstream response, usually activation of an intermediate enzyme or production of a second messenger in a signalling pathway. These include the G-protein-coupled, tyrosine kinase and nuclear receptors. Other receptors are actually ion channels, which may be triggered either by ligand binding or a change in membrane voltage. These trigger cellular responses by changing the distribution of ions, for example Na^+, K^+, and Ca^{2+}, between the inside and outside of the cell.

4.3 Ion channels may be ligand- or voltage-gated, where opening of the channel will allow the transfer of ions either in or out of the cell. Nicorandil is a K^+ channel opener, which allows K^+ ions to leave the cell. This causes a net hyperpolarization (negative charge) of the plasma membrane. The downstream effect of this is that the nearby Ca^{2+} channels, which are voltage-gated and therefore sensitive to changes in the potential difference across the plasma membrane, fail to open. The biological response is that Ca^{2+} ions will not be able to flow into vascular smooth muscle cells and initiate contraction. This causes vasodilatation of the coronary arteries, increasing blood flow to support cardiac contraction in angina patients. This is an example of how inhibition of one ion channel may initiate changes in downstream ion channels to elicit a biological response.

Another example is found in the way the cardiac glycoside digoxin works as an inotropic agent (increases the force of contraction or contractility). Digoxin binds to the cardiac glycoside receptor on the Na$^+$/K$^+$ exchange pump, which normally pumps K$^+$ ions into the cell and Na$^+$ out of the cell. If this is prevented by digoxin, however, Na$^+$ ions will accumulate inside the cell in the cytoplasm. Another ion exchange pump, which exchanges Na$^+$ for Ca^{2+}, depends upon there being a concentration gradient such that there is a higher concentration of Na$^+$ ions outside the cell than in the cytoplasm. If this is so, Na$^+$ ions may be pumped back into the cell in exchange for an outward flow of Ca^{2+} ions. If, however, there is already an accumulation of Na$^+$ ions inside the cell, the concentration gradient will not allow this exchange to take place, causing an overall increased intracellular Ca^{2+} and therefore an increased force of contraction of the muscle.

4.4 Efficacy is the relationship between receptor binding and the pharmacodynamic response, where E$_{max}$ is the maximum biological response produced when all receptor binding sites are filled. Potency, however, is the concentration of drug that produces a half maximal response (EC$_{50}$). Efficacy and E$_{max}$ are measures of the intensity of the response and therefore may be used preclinically to predict how well a drug will treat the symptoms or causes of a disease, that is, may translate into clinical effectiveness. Potency, however, may be more closely linked to the dosage of a drug.

4.5 Certain drugs, such as cocaine, amphetamines, diamorphine, and cannabis, are highly addictive because they cause feelings of euphoria or a 'high' when they trigger certain neurological reward pathways. These pathways involve two major neurotransmitters, serotonin and dopamine, linked to the mesocorticolimbic system or 'pleasure centre' of the brain. Physiological dependence might also be accompanied by psychological dependence, where drug misusers take drugs repeatedly for personal satisfaction.

Chapter 5

5.1 Nervous signals can travel from one end of the body to the other in milliseconds, and can often be discretely targeted to one muscle or organ for a defined period of time.

5.2 The myelin sheath around axons in the peripheral nervous system is formed by Schwann cells, which are a type of neuroglial cell.

5.3 Action potential propagation is carried out by voltage-gated sodium channels.

5.4 Myelination allows saltatory conduction, whereby the active zone 'jumps' between nodes of Ranvier, speeding up action potential conduction. Larger diameter nerve fibres have a lower internal electrical resistance than smaller diameter fibres, and conduction velocity is therefore greater.

5.5 Aα motor neurons conduct action potentials the fastest (up to 120 m/s), as they are large diameter myelinated fibres.

5.6 Neurotransmitters must act to transmit discrete signals, that is, transmit each action potential separately. The delay between action potentials may be very short in rapidly firing nerves (such as motor neurons), and neurotransmitters must be removed before the next action potential

reaches the nerve terminal. If neurotransmitters are not removed quickly, depolarizing block may occur.

5.7 Excitation–contraction coupling is the conversion of an electrical signal (action potential) into a muscle contraction. The action potential travelling along T-tubules in a muscle cell activates voltage-gated calcium channels. The rise in calcium signals the contractile elements of a muscle fibre (actin and myosin) to interact, and the muscle contracts.

5.8 Sensory receptor activation (for example, stimulation by light, mechanical touch, etc.) leads to a change in membrane potential called the receptor potential. If this receptor potential is large enough, an action potential will be generated.

5.9 Local anaesthetics work by blocking action potential propagation along neuronal axons. They do this by physically blocking voltage-gated sodium channels, which are responsible for action potential propagation.

5.10 The two mechanisms for neuromuscular block are depolarizing block and non-depolarizing block. Depolarizing block is caused by initial depolarization of the postsynaptic membrane, followed by desensitization and antagonism of neurotransmitter binding. Non-depolarizing block is caused by antagonism of the nicotinic acetylcholine receptor, preventing neurotransmitter binding.

Chapter 6

6.1 Drugs which increase the neurotransmitter indirectly are termed indirect acting. For example, drugs that inhibit the reuptake of noradrenaline (sympathetic) or block acetylcholinesterase (parasympathetic) will consequently increase the concentration of the neurotransmitter at the synapse.

6.2 Increased release and/or reduced reuptake of noradrenaline would occur throughout the nervous system, causing a massive elevation in peripheral and central noradrenaline, producing potentially life-threatening effects on blood pressure and the cardiovascular system, as well as complex CNS effects.

6.3 Their widespread actions and extensive non-specific effects make ganglionic stimulators and blockers unsuitable for clinical use because nicotinic receptors are located in many places throughout the nervous system. The most obvious nicotinic stimulant in clinical use is **nicotine**, used in nicotine withdrawal therapy to gradually wean smokers off cigarettes. Excessive smokers only may experience autonomic effects on the gastrointestinal tract and increased secretions, as very high levels are needed to create these effects.

6.4 α-Adrenoceptor expression predominates in blood vessels supplying the salivary glands, producing reduced secretions during the 'flight or fight' responses, whereas in the blood vessels supplying skeletal muscle, β$_2$-adrenoceptor expression predominates to provide oxygen to muscles being used in these responses.

6.5 The increase in pressure difference will cause an increase in flow rate in the periphery, thereby increasing tissue perfusion.

6.6 Only the peripheral nervous system utilizes noradrenaline as a neurotransmitter and adrenaline as a hormone. Some degree of selectivity, and thus clinical efficacy, can therefore be achieved simply by inhibiting noradrenergic neurotransmission, which is not possible with cholinergic neurotransmission. The discovery that different receptor subtypes mediate different aspects of autonomic function greatly facilitated the development of effective therapeutic agents with reduced side-effect profiles.

Chapter 7

7.1 Histamine has two principal effects on peripheral blood vessels. First, histamine promotes the dilation of small blood vessels, thus increasing the amount of blood contained within them. Second, it promotes an increase in the size of the small gaps between adjacent endothelial cells. This has the effect of allowing small to medium-sized proteins to leak out into the surrounding tissue space. Because of the affinity the proteins have for water there will also be a net movement of water from the blood into the tissues, causing an accumulation of fluid outside the blood vessels. This is known as oedema.

7.2 Hay fever is a seasonal allergic condition that is triggered by wind-borne pollen. It may be treated in a number of different ways but its symptoms are often controlled with antihistamine drugs that are competitive antagonists for histamine at the H_1 receptor. Some antihistamines are short acting and have to be administered several times a day. These drugs generally achieve effective plasma levels very quickly and are useful for treating symptoms. Others are much longer acting but may take several days to achieve effective levels in the blood. These drugs are valuable for daily treatment throughout the season.

7.3 Headaches are often associated with dilation of the blood vessels of the brain. Stimulation of $5-HT_1$ receptors, which results in vasoconstriction of these vessels, can alleviate this problem.

7.4 TXA_2 promotes vasoconstriction and the aggregation of platelets, which together are very valuable responses to a damaged blood vessel. PGI_2 promotes local vasodilatation and inhibits platelet adhesion to the inside lining of blood vessels. The ability of PGI_2 to counter the actions of TXA_2 is a good example of different eicosanoids exerting opposing effects. This allows moderation at a very local level.

7.5 The theoretically ideal dose of aspirin would be one that would inhibit thromboxane synthesis by platelets without inhibiting the synthesis of PGI_2 by endothelial lining cells of blood vessels. There is evidence that a very low dose of aspirin (75 mg) can achieve this differential inhibition.

7.6 Diabetes is a loss of control of glucose levels. Normally, the body controls the level of glucose in the body within very narrow limits. Following a meal, blood glucose levels rise as nutrients are absorbed from food. The body responds to this by secreting insulin from the pancreas. Insulin stimulates the movement of glucose out of the blood into storage sites, principally the liver and skeletal muscles. In between meals, when blood glucose reaches a low level, the body releases glucagon from the pancreas, which promotes the release of glucose from storage in the liver.

In diabetics this control is lost, either because the subject has lost the ability to produce insulin (type 1, usually occurring in children) or because the body has become unresponsive to insulin (type 2, usually occurring in adults).

Type 1 diabetes is usually controlled by injecting insulin at appropriate times. Type 2 diabetes is usually controlled by limiting the amount of carbohydrate in the diet or using hypoglycaemic drugs such as metformin or glibenclamide.

Chapter 8

8.1 Homeostasis is the process by which an organism maintains its internal environment. It is important that physiological factors are kept within well-defined limits or illness and death will occur.

8.2 Negative feedback describes the situation whereby the products of a system inhibit their own production. Negative feedback mechanisms are important in homeostasis because they tend to lead to stable systems.

8.3 Redundancy is displayed by a system which uses more than one method of varying a single factor. It is helpful in homeostasis because it means that a system can make finer adjustments, and compensation can be made for parts of the system that are not functioning correctly.

8.4 Endotherms control their own body temperature, which is generally higher than that of ectotherms. The body temperature of ectotherms depends largely on the ambient temperature. Mammals are endotherms, reptiles are ectotherms.

8.5 When the core temperature is too high, the body initiates a number of mechanisms to increase dissipation of heat. These include: sweating (in which heat is dissipated by the evaporation of water from the skin), peripheral vasodilatation (to enable the blood to flow closer to the skin surface, so heat is more easily lost), and flattening of skin hair by the relaxation of arrector pili (which reduces the trapping of air, a thermal insulator, close to the skin).

8.6 When the body temperature is too low, the body initiates a number of mechanism to increase heat production, and to reduce heat loss from the body. These include: Shivering (which increases the metabolic rate in muscles, generating heat as a by-product), non-shivering thermogenesis (generation of heat in brown adipose tissue), peripheral vasoconstriction (which redirects some blood flow away from the skin, into deeper vessels where thermal insulation is better), and causing of the body hair to stand upright (which traps a layer of insulating air close to the body).

8.7 Hypotension (low blood pressure) leads to dizziness. If it is severe it may cause muscle weakness and fainting. Hypertension (high blood pressure) damages all organs and tissues. Most seriously it increases the risk of myocardial infarction and stroke.

8.8 Hypoglycaemia causes sweating, irritability, and hunger. In severe cases it can lead to a lack of coordination and eventually coma. Hyperglycaemia, when prolonged, damages an array of tissues by glycosylation, leading to retinopathy and nephropathy. Very severe hyperglycaemia is associated with diabetic ketoacidosis.

8.9 Calcitonin lowers plasma calcium concentration by inhibiting the bone reabsorbing actions of osteoclasts and promoting renal calcium secretion. Parathyroid hormone increases plasma calcium concentration by promoting osteoclast activity and inhibiting renal reabsorption of calcium. 1,25-Dihydrocycholecalciferol maximizes calcium uptake from the intestines.

Chapter 9

9.1 Red blood cells (erythrocytes), white blood cells (leukocytes), and platelets.

9.2 Erythropoietin is the hormone secreted by the kidneys that stimulates erythropoiesis (RBC production).

9.3 The primary function of haemoglobin is to bind and transport oxygen around the body.

9.4 WBCs are formed in the bone marrow via a process known as leukopoiesis. The five types of WBC are: basophils, eosinophils, lymphocytes, monocytes, and neutrophils.

9.5 The substances found in plasma are: albumin; hormones; lipids; antibodies; fibrinogen; coagulation factors; enzymes; coenzymes; sodium, potassium, magnesium, calcium, chloride, bicarbonate, sulphate, and ammonium ions; cholesterol; carbohydrates; amino acids; urea; uric acid; creatine; and bilirubin.

9.6 Anaemia is a condition characterized by a deficiency in RBCs whereby there is a reduction in the oxygen-carrying capacity of blood. The symptoms include: tiredness, weakness, dizzy spells, and feeling faint.

9.7 Leukaemia is a term used to define a broad range of cancers that affect the functions of WBCs. There are two main types of acute leukaemia: acute lymphoblastic leukaemia (ALL) and acute myeloid leukaemia (AML). The main types of chronic leukaemia are chronic lymphocytic leukaemia (CLL) and chronic myeloid leukaemia (CML).

9.8 The four phases of haemostasis are:
1. Vascular phase
2. Platelet phase
3. Coagulation phase
4. Fibrinolysis

Platelets initiate the formation of a fibrin clot by covering the damaged area of a blood vessel, to prevent further blood loss.

9.9 See Table below.

Recipient	Donor							
	O−	**O+**	**A−**	**A+**	**B−**	**B+**	**AB−**	**AB+**
O−	Yes	No	No	No	No	No	No	No
O+	Yes	Yes	No	No	No	No	No	No
A−	Yes	No	Yes	No	No	No	No	No
A+	Yes	Yes	Yes	Yes	No	No	No	No
B−	Yes	No	No	No	Yes	No	No	No
B+	Yes	Yes	No	No	Yes	Yes	No	No
AB−	Yes	No	Yes	No	Yes	No	Yes	No
AB+	Yes	Yes	Yes	Yes	Yes	Yes	Yes	Yes

Index